Cheminformatics and its Applications

Edited by Amalia Stefaniu, Azhar Rasul and Ghulam Hussain

Published in London, United Kingdom

IntechOpen

Supporting open minds since 2005

Cheminformatics and its Applications
http://dx.doi.org/10.5772/intechopen.83236
Edited by Amalia Stefaniu, Azhar Rasul and Ghulam Hussain

Contributors
Dionisio Antonio Olmedo A., José Luis Medina-Franco, João D Ferreira, Francisco M Couto, Daniel Glossman-Mitnik, Norma Flores-Holguín, Juan Frau, José Ciriaco-Pinheiro, Heriberto Bitencourt, José Lobato, Antonio Florêncio De Figueiredo, Marcos Antonio Dos Santos, Fabio Gil, Raimundo Ferreira, Luã Felipe De Oliveira, Sady Alves, Edilson Luiz C De Aquino, Márcio De Souza Farias, Yu-Chen Lo, Hiroshi Honda, Gui Ren, Azhar Rasul, Ammara Riaz, Iqra Sarfraz, Ayesha Sadiqa, Javaria Nawaz, Rabia Zara, Samreen Gul Khan, Zeliha Selamoglu, Wolfgang Fecke, Bahne Stechmann, Kenji Sorimachi, Sonia Aroui, Amalia Stefaniu, Kara L. Davis, Abderraouf Kenani

First published in London, United Kingdom, 2020 by IntechOpen
IntechOpen is the global imprint of INTECHOPEN LIMITED, registered in England and Wales, registration number: 11086078, 7th floor, 10 Lower Thames Street, London, EC3R 6AF, United Kingdom
Printed in Croatia

British Library Cataloguing-in-Publication Data
A catalogue record for this book is available from the British Library

Additional hard and PDF copies can be obtained from orders@intechopen.com

Cheminformatics and its Applications
Edited by Amalia Stefaniu, Azhar Rasul and Ghulam Hussain
p. cm.
Print ISBN 978-1-83880-067-3
Online ISBN 978-1-83880-068-0
eBook (PDF) ISBN 978-1-83962-518-3

For EU product safety concerns:
IN TECH d.o.o., Prolaz Marije Krucifikse Kozulić 3, 51000 Rijeka, Croatia, info@intechopen.com or visit our website at intechopen.com.

Meet the editors

Amalia Stefaniu has a background in chemical engineering, acquiring her Bachelor's degree at Politehnica University of Bucharest, Faculty of Engineering in Foreign Languages. She followed postgraduate academic studies with *Drugs and Cosmetics* specialization and she obtained a Masters degree in *Biotechnologies and food safety*. She completed her PhD in Exact Sciences - Chemistry Domain in 2011 at the University Politehnica of Bucharest, Faculty of Applied Chemistry and Materials Science, Department of Inorganic Chemistry, Physical Chemistry and Electrochemistry. She joined the National Institute for Chemical Pharmaceutical Research and Development, Bucharest in 2001, where she worked first as Chemical Research Engineer in the Pharmaceutical Biotechnologies domain. Her current position is Senior Research Scientist. Her research focuses on properties prediction, mathematical modeling, molecular docking, and therapeutic compounds design.

Dr. Azhar Rasul is an Assistant Professor at the Government College University, Faisalabad. He obtained his PhD fellowship jointly awarded by MoE, Pakistan and CSC, China and completed his Ph.D. in Chemical Cancer Biology from the Northeast Normal University, China. He later received China Postdoctoral Fellowship in 2012, Japanese Society for Promotion of Science (JSPS) Postdoctoral Fellowship in 2013, and subsequently Tokyo Biochemical Research Foundation (TBRF) Fellowship in 2015. He has published over 100 peer-reviewed articles with a cumulative impact factor over 208 and with over 1670 citations. He has presented several invited talks at the national and international level. He has obtained several national and international research grants. His laboratory is actively engaged in interdisciplinary research on novel tumor biomarkers and identification of non-toxic anti-cancer compounds for various hallmarks of cancer from natural sources. He is a reviewer and editorial board member of several well-reputed journals.

Dr. Ghulam Hussain is working as an Assistant Professor at the Department of Physiology, Government College University, Faisalabad. Dr. Hussain has served as a visiting scientist at Huaqiao University, Xiamen, China. He earned his MPhil and Ph.D. in Neurosciences from the University of Strasbourg, France under the Overseas Scholarship Program of Higher Education Commission of Pakistan. He has published 70 peer-reviewed articles with a cumulative impact factor of 120 and 480 citations. He has presented his work at both the national and international level. He is a recipient of two research grants from HEC Pakistan. He is also working as a reviewer for well-reputed research journals. His laboratory is involved in elucidating the possibilities of promoting peripheral nerve regeneration following traumatic injury.

Contents

Preface

Cheminformatics has emerged as an applied branch of Chemistry that involves multidisciplinary knowledge, connecting related fields such as chemistry, computer science, biology, pharmacology, physics, and mathematical statistics. Computational methods are used to visualize simple structures or macromolecular assemblies, to model properties by mathematical and statistical models, to create, store and process chemical data (databases, data mining), to realize virtual screening of large compound libraries and to analyze the chemical information and optimize structure in order to develop novel compounds, materials, or processes.

The book is organized in two sections, covering plural aspects related to advances in the development of informatic tools and their specific use in compound databases and concerted efforts to link them in research platforms and networks with various purposes and applications in life sciences. Applications in medicinal chemistry, for identification and development of new therapeutically active molecules are described, but the book is not limited to these topics. For instance, the chapter titled "Visible Evolution from Primitive Organisms to Homo sapiens" covers the area of genomic analysis and development of evolutionary equations based on genome structure. It represents an important approach to explain the origin and evolution of life, providing mathematical proofs on the genomic amino acid composition homogeneity. It illustrates the use of mathematics to explain biological organisms' evolution and reduces complex structural genetic information to simple linear regression relationships. This chapter allows inexperienced readers to understand the basic concepts and theory, but also invites them to go forward, offering deep biological and chemical molecular insights.

The chapter titled "Semantic similarity in cheminformatics" presents a great overview of chemical ontologies, explaining how it works, how the relationships between different chemical or biological entities are constructed in order to bind chemical information given by structures with other aspects as chemical classifications, reaction mechanisms, metabolites, toxicity, biological pathways and so on. The authors describe the fundamental concepts of ontology-based semantic similarity, pointing to the applications in cheminformatics and discussing the efforts in ontology development to link chemical databases with related fields such as medical chemistry, genomics, or proteomics.

Computational tools of chemometrics and pattern recognition techniques are used for the design of various compounds. Such examples are illustrated in the chapter titled "Molecular Electrostatic Potential and Chemometric Techniques as Tools to Design of Bioactive Compounds", where authors use *ab initio* calculation of properties based on charge density and topological indices for the design of nitrofurans derivatives. The key features and descriptors, acting in the recognition process with the biological target, are elucidated and can be further used to design new biologically active molecules.

The next chapter ("Chemical reactivity properties and bioactivity scores of the Angiotensin II vasoconstrictor octapeptide") emphasizes the reactivity descriptors, drug-likeness assessment, and prediction of oral bioavailability scores as preliminary steps for the development of new drugs based on specific peptide analogues, achieving a comparison of prediction realized with different quantum mechanical modelling methods.

Molecular complexity, flexibility, and other structural features and properties are used in a cheminformatic analysis of natural and synthetic compounds, based on similarity, in a case study of products originating from Panama, in an attempt to find and optimize lead compounds with antimalarial activity, in the chapter "Cheminformatic Approach: The Case of Natural Products of Panama".

In the chapter titled "Accelerating chemical tool discovery by academic collaborative models", the authors highlight the international efforts of academia and industrial pharmacists to generate consortia in the interdisciplinary field of chemical biology, to connect their knowledge, compound libraries and facilities, having the important goal to create open access information. The principal aim remains the development of new therapeutic compounds using the knowledge from multidisciplinary fields in academic and public and private media, thus helping researchers to solve mechanistical issues in life sciences.

The chapter "Chemical Biology Toolsets for Drug Discovery and Target Identification" is an overview of chemical techniques and methodologies implemented in the study of biological systems, metabolic pathways, drug-target complex interactions, and other biochemical process, all with the common goal to understand the action and all biochemical implications of the introduction in therapeutics of a new drug. Different complementary instrumental techniques and methodologies aiming to provide deep insights into the chemical structure are discussed alongside validation methods and techniques of selection of a new drug candidate.

Machine learning and deep learning are aspects covered in the chapter titled "Machine-learning based drug discovery and design", presenting a detailed view of their theoretical aspects and applications related to *de novo* drug design, QSAR analysis, and chemical space visualization

The chapter titled "Cell Penetrating Peptides", as its title suggests, emphasizes their biomedical applications as transport vectors for different therapeutic agents across cell membranes. The authors describe the origin and the classifications of CPPs, their uptake mechanisms, and their promising clinical efficacity in various cancer therapies.

With all information and conclusive examples presented above, this book is a valuable learning resource for readers from the scientific community, students, researchers both beginners and experienced in the field of chemistry/bioinformatics and related domains. By taking note of these chapters, I hope readers will feel encouraged, inspired, and motivated to continue new research and discoveries.

I thank all authors for their substantial contributions to this book, for sharing their knowledge, and for opening new opportunities and perspectives in such an evolving field as cheminformatics is.

Amalia Stefaniu
National Institute for Chemical - Pharmaceutical Research and Development – ICCF Bucharest (Romania),
Department of Pharmaceutical Biotechnologies,
Laboratory of Molecular Design and Molecular Docking,
Bucharest, Romania

Dr. Azhar Rasul
Department of Zoology,
Faculty of Life Sciences,
Government College University Faisalabad (GCUF),
Faisalabad, Pakistan

Ghulam Hussain
Department of Physiology,
Faculty of Life Sciences,
Government College University Faisalabad (GCUF),
Faisalabad, Pakistan

Section 1

Insights of Chemical Structures by Chemoinformatics Approaches

Chapter 1

Prologue: Deep Insights of Chemical Structures by Chemoinformatics Tools, Let's Think Forward!

Amalia Stefaniu

1. Introduction - Multidisciplinary context

The constant need of chemical scientists to understand complex phenomena and process and to achieve a rational structural design by controlling the synthesis to obtain compounds with improved properties or materials with enhanced quality, together with advances in information technology, has led to development of a new branch of chemistry—chemoinformatics—with strong implications in life sciences such as molecular biology or biochemistry, with major interest in medicine, pharmaceutical and food science industries.

Mainly, these interdisciplinary efforts are focused on the medical and pharmaceutical area, aiming to improve the quality and standard of life, and have applications in drug design and development of new therapeutic strategies. Chemoinformatics, as new discipline, covers a broad spectrum of aspects including all applications of information technology to chemistry involving: constructing and archiving big compound libraries (small molecules and proteins) containing structural properties and molecular descriptors, spectra, X-ray crystallography data and so on; information processing; large-scale chemical data mining; computational tools for structure and interactions visualisation, computational models for predicting interactions, to calculate properties and bioactivity, molecular docking and dynamic simulations methodologies, virtual screening, pharmacophore modelling, fragments similarity analysis, estimation of ADME (absorption, distribution, metabolism and excretion) characteristics, toxicity alerting, etc. [1–4]. The integration of chemical information and its transformation involves mathematical models and statistical data analysis.

Due to web servers and open data initiatives, large amount of chemical data from screening libraries are now available [5] and facilitate the drug discovery process. There are numerous chemoinformatics databases which contain various experimental and/or predicted properties of small molecules (ligands), peptides, proteins and data about their interactions (drug-drug interactions, ligand-protein interactions, protein-protein interactions, RNA-ligand interactions), chemical toxicity, bioactivity, adverse drug reactions, drug pathways, toxicogenomics, secondary metabolites, pharmacokinetics, etc. The existing data could help to build new structures and new models and to make new in silico predictions about physico-chemical properties and behaviour.

To raise awareness of the outstanding importance and impact of chemoinformatics research, exemplified below are some of its applications in life sciences, preponderant in medicinal chemistry.

2. Applications of chemoinformatics in medicinal chemistry

Novel druggable protein targets are a subject of research in order to develop new therapeutic strategies against various diseases (scleroderma, Alzheimer's disease, infections, etc.). Investigations include methods such as quantitative structure-activity relationships (QSAR), similarity search, pharmacophore modelling, molecular docking and dynamic simulations and toxicity assessment.

2.1 Anticancer therapy design

To fight against malignancies, new screening methods aim to identify and develop novel chemical antiproliferative agents, with promising results. As example, biomolecular modelling techniques are used to identify potential kinase inhibitor targets. The mitogen-activated protein kinase (MAPK) plays a key role in tumorigenesis; that is why it is considered a priority druggable target candidate for anticancer therapy. The interactions of cancer-related MAPK kinases and potential inhibitors are investigated by in silico tools. Molecular docking calculations are employed to predict the inhibitor-bound active sites and the binding modes for actual and potential anticancer drugs [6].

2.2 Parkinson's disease

Researchers' efforts to improve medication for Parkinson's disease benefit from chemoinformatics and molecular docking tools to identify new potential neuroprotective compounds able to effectively treat the disease, by inhibition of oligomerization process of α-synuclein protein. By computational techniques, the protein in its dimer and oligomer forms can be studied, and multiple molecules are subject of computational simulations in order to identify potential inhibitors of α-synuclein aggregation [7].

2.3 Alzheimer's disease

Chemoinformatics approaches including molecular docking, dynamic simulations, lead optimization and quantum chemical characterisation are used to achieve the inhibition of acetylcholinesterase (AChE) and butyrylcholinesterase (BChE) enzymes, responsible for cholinergic dysfunctions associated with the cognitive and behavioural abnormalities in dementing illness, in order to design and develop new therapeutic agents against this disease [8–11]. Other approaches focus on the amyloid-beta aggregation process, trying to stop the formation of neurotoxic species, and the design of new inhibitors, the study being also facilitated by computational techniques such as QSAR modelling and assessment of inhibition efficiency by predicting stability and binding modes of potential inhibitors through combined computational techniques including structure-activity relationships analysis, docking and molecular dynamic simulations [12–15].

2.4 Antimicrobial agents

Researchers focus their studies to block the activity of DNA gyrase and topoisomerase IV, which are essential bacterial enzymes involved in replication and recombination processes. The design of novel antibacterial agents that act against these enzymes can be realised by molecular docking techniques and bioactivity evaluation. That is the case of quinolones, which act equally against DNA gyrase and topoisomerase IV [16–19].

Pharmacokinetics/ADMET properties such as absorption, distribution, metabolism, excretion and toxicity of designed structures are assessed through computational approaches too, aiming to predict the therapeutic potential of the lead compound. Biochemical properties and drug-likeness according Lipinski's rule of five (RO5) [20] and the molecular flexibility, as key descriptors to describe the oral bioavailability of drugs, are also predicted using computational tools. Thus, computer-aided drug design, coupled with in silico ADMET studies, helps to select the drug candidate molecules with possible better efficacy and less side effects (poor hepatotoxic effects).

3. Application in identification and quantification of substances of abuse

Recent researches report the application of chemometric tools in correlation with spectrometric techniques (near-infrared spectroscopy) for onsite analysis of cannabinoids or amphetamine compounds (with portable and handheld NIR devices). The chemometric tools allow the user to compare collection of spectra, to develop prediction models and to achieve a real-time detection of sample contamination. Such method could become an alternative way of detection of illicit drugs, determined in oral fluids, being non-invasive, rapid and accurate test, completely automated [21, 22].

4. Applications in food chemistry

Food chemical data sets can be manipulated and analysed also by computational resources similar with those for drugs and nutraceuticals. The interest in this area is growing because of the food-related industrial challenges. Thus, an emerging field of research has arisen: foodinformatics [23]. In silico quantitative approaches are used to assess genotoxicity and carcinogenicity of food additives (flavours, colourants, contaminants, etc.) or cosmetic ingredients [24–26], in the attempts of safety evaluation for the human health. All these computational approaches must be verified by in vitro methods.

This section is a collection of advanced studies focusing on topics of interest in the context of chemoinformatics applications in drug discovery and design of new molecules.

Author details

Amalia Stefaniu
Department of Pharmaceutical Biotechnologies, National Institute of Chemical Pharmaceutical Research and Development (ICCF), Bucharest, Romania

*Address all correspondence to: astefaniu@gmail.com

References

[1] Vogt M, Bajorath J. Chemoinformatics: A view of the field and current trends in method development. Bioorganic & Medicinal Chemistry. 2012;**20**:5317-5323

[2] Begam BF, Kumar JS. A study on Cheminformatics and its applications on modern drug discovery, international conference on modeling optimisation and computing (ICMOC 2012). Procedia Engineering. 2012;**38**:1264-1275

[3] Lo Y-C, Rensi SE, Torng W, Altman RB. Machine learning in chemoinformatics and drug discovery. Drug Discovery Today. 2018;**23**(8):1538-1546

[4] Gasteiger J. Chemoinformatics: Achievements and challenges, a personal view. Molecules. 2016;**21**(2):151. DOI: 10.3390/molecules21020151

[5] Gonzalez-Medina M, Naveja JJ, Sanchez-Cruz N, Medina-Franco JL. Open chemoinformatic resources to explore the structure, properties and chemical space of molecules. RSC Advances. 2017;7:54153. DOI: 10.1039/c7ra11831g

[6] Meng L, Huang Z. In silico-in vitro discovery of untargeted kinase–inhibitor interactions from kinase-targeted therapies: A case study on the cancer MAPK signalling pathway. Computational Biology and Chemistry. 2018;**75**:196-204

[7] Rondon-Villarreal P, Lopez WOC. Identification of potential natural neuroprotective molecules for Parkinson's disease by using chemometrics and molecular docking. Journal of Molecular Graphics and Modelling. 2020;**97**:107547

[8] Hassan M, Abbasi MA, Aziz-Ur-Rehaman, Siddiqui SZ, Hussain G, Shah SAA, et al. Exploration of synthetic multifunctional amides as new therapeutic agents for Alzheimer's disease through enzyme inhibition, chemoinformatic properties, molecular docking and dynamic simulation insights. Journal of Theoretical Biology. 2018;**458**:169-183

[9] Makhaeva GF, Kovaleva NV, Boltneva NP, Lushchekina SV, Rudakova EV, Stupina TS, et al. Conjugates of tacrine and 1,2,4-thiadiazole derivatives as new potential multifunctional agents for Alzheimer's disease treatment: Synthesis, quantum-chemical characterization, molecular docking, and biological evaluation. Bioorganic Chemistry. 2020;**94**:103387

[10] Hassan M, Abbasi MA, Aziz-Ur-Rehaman, Siddiqui SZ, Shahzadi S, Raza H, et al. Designing of promising medicinal scaffolds for Alzheimer's disease through enzyme inhibition, lead optimization, molecular docking and dynamic simulation approaches. Bioorganic Chemistry. 2019;**91**:103138

[11] Dhanjal JK, Sharma S, Grover A, Das A. Use of ligand-based pharmacophore modeling and docking approach to find novel acetylcholinesterase inhibitors for treating Alzheimer's. Biomedicine & Pharmacotherapy. 2015;**71**:146-152

[12] Safarizadeh H, Garkani-Nejad Z. Molecular docking, molecular dynamics simulations and QSAR studies on some of 2-arylethenylquinoline derivatives for inhibition of Alzheimer's amyloid-beta aggregation: Insight into mechanism of interactions and parameters for design of new inhibitors. Journal of Molecular Graphics and Modelling. 2019;**87**:129-143

[13] Eskici G, Gur M. Computational design of new peptide inhibitors for amyloid beta (Aβ) aggregation in Alzheimer's disease: Application of a novel methodology. PLOS One.

2013;**8**(6):e66178. DOI: 10.1371/journal.pone.0066178

[14] Tran L, Kaffy J, Ongeri S, Ha-Duong T. Binding modes of a glycopeptidomimetic molecule on Aβ protofibrils: Implication for its inhibition mechanism. ACS Chemical Neuroscience. 2018;**9**(11):2859-2869. DOI: 10.1021/acschemneuro.8b00341

[15] Tonali N, Kaffy J, Soulier JL, Gelmi ML, Erba E, Taverna M, et al. Structure-activity relationships of β-hairpin mimics as modulators of amyloid β-peptide aggregation. European Journal of Medicinal Chemistry. 2018;**154**:280-293. DOI: 10.1016/j.ejmech.2018.05.018

[16] Pintilie L, Stefaniu A, Nicu AI, Caproiu MT, Maganu M. Synthesis, antimicrobial activity and docking studies of novel 8-chloro-quinolones. Revista de Chimie (Bucharest). 2016;**67**(3):438-445

[17] Pintilie L, Stefaniu A, Nicu AI, Maganu M, Caproiu MT. Design, synthesis and docking studies of some novel fluoroquinolone compounds with antibacterial activity. Revista de Chimie (Bucharest). 2018;**69**(4):815-822

[18] Strahilevitz J, Hooper DC. Dual targeting of topoisomerase IV and Gyrase to reduce mutant selection: Direct testing of the paradigm by using WCK-1734, a new fluoroquinolone, and ciprofloxacin. Antimicrobial Agents and Chemotherapy. 2005;**49**(5):1949-1956

[19] Collin F, Karkare S, Maxwell A. Exploiting bacterial DNA gyrase as a drug target: Current state and perspectives. Applied Microbiology and Biotechnology. 2011;**92**:479-497. DOI: 10.1007/s00253-011-3557-z

[20] Lipinski CA, Lombardo F, Dominy BW, Feeney PJ. Experimental and computational approaches to estimate solubility and permeability in drug discovery and development settings. Advanced Drug Delivery Reviews. 2001;**46**:3-26

[21] Risoluti R, Gullifa G, Buiarelli F, Materazzi S. Real time detection of amphetamine in oral fluids by MicroNIR/Chemometrics. Talanta. 2020;**208**:120456

[22] Risoluti R, Gullifa G, Battistini A, Materazzi S. Monitoring of cannabinoids in hemp flours by MicroNIR/Chemometrics. Talanta. 2020;**211**:120672

[23] Martinez-Mayorga K, Medina-Franco JL, editors. Foodinformatics: Applications of Chemical Information to Food Chemistry. 2014th ed. Cham, Heidelberg, New York, Dordrecht, London: Springer. ISBN 978-3-319-10225-2; ISBN 978-3-319-10226-9 (eBook). DOI 10.1007/978-3-319-10226-9

[24] Tcheremenskaia O, Battistelli CL, Giuliani A, Benigni R, Bossa C. In silico approaches for prediction of genotoxic and carcinogenic potential of cosmetic ingredients. Computational Toxicology. 2019;**11**:91-10

[25] Benigni R. Towards quantitative read across: Prediction of Ames mutagenicity in a large database. Regulatory Toxicology and Pharmacology. 2019;**108**:104434

[26] Kruhlak NL, Contrera JF, Benz RD, Matthews EJ. Progress in QSAR toxicity screening of pharmaceutical impurities and other FDA regulated products. Advanced Drug Delivery Reviews. 2007;**59**:43-55

Chapter 2

Visible Evolution from Primitive Organisms to *Homo sapiens*

Kenji Sorimachi

Abstract

The ratios of amino acids to the total amino acids deduced from the complete genome and those of nucleotides to the total nucleotides in the genome are useful indexes to characterize various large genomes among different species from bacteria to *Homo sapiens*. These indexes are not only independent of species but also of genome size. Using these indexes, the following results were obtained: (1) primitive life forms appeared to have similar amino acid compositions to present day organisms; (2) cellular amino acid compositions that are similar among various species and between whole cells and complete genomes; (3) genome structure that is homogeneously constructed from putative small units encoding proteins of similar amino acid compositions, followed by synchronous mutations over the genome; (4) all organisms can be classified into two groups, "GC-rich" and "AT-rich," based on their nucleotide contents, or "terrestrial" and "aquatic vertebrates" based on natural selection by cluster analyses using amino acid contents as the traits; and (5) evolution based on nucleotide content alterations can be expressed by definitive equations. Thus, the ratios of amino acids or nucleotides to their total contents are useful indexes for characterizing genomes, regardless of species differences and genome sizes. The two normalized nucleotide contents are universally expressed regression line.

Keywords: genome, mitochondria, codons, Chargaff's parity rules, cluster analysis, normalization, phylogenetic trees, evolution

1. Introduction

The origin of life has long been interested to human since old times. Indeed, Aristotle proposed "spontaneous generation" more than 2000 years ago, although this idea was disproved by Louis Pasteur in experiments using "swan neck flasks." Our great interest in the origin of life might be expressed by the following philosophical words: *Where do we come from? What are we? Where are we going?* These words were written by French artist Paul Gauguin on his canvas in Tahiti in 1897.

The development of nucleotide sequencing technology [1, 2] has contributed to progress in molecular biology, including the analysis of a complete bacterial genome first carried out in 1995 [3], and, subsequently, the draft human genome, which was reported in 2001 [4, 5]. At present (June 19, 2019), 498 eukaryote, 5159 bacterial, and 296 archaeal complete genomes were determined. However, the origin of life is still unclear. Assuming that the replacement rates of nucleotides or amino acids in genes are constant [6], phylogenetic trees were drawn [6–11]. However, we know that their exact replacement rates differ between genes and between species. Studies based on nucleotide or amino acid sequences are applicable to genes whose nucleotide or amino acid numbers are much smaller than those of complete genomes, but not to genomes consisting of huge numbers of nucleotides and many genes. Of course, simple comparison of sequence differences between genes in the same species and the same genes in different species is useful.

2. Normalization

Intraspecies nucleotide contents were first analyzed in 1950 by Chargaff, who reported that G = C, A = T, and [(G + A) = (C + T)] [12], which was named as Chargaff's first parity rule. This rule is understandable based on the double-stranded DNA structure [13]. Additionally, this rule is applicable to single-stranded DNA obtained from a single species nucleus, termed Chargaff's second parity rule [14]. As the rules are based on normalized values to 1 (G + C + A + T = 1), nucleotide contents are expressed by their ratios. However, the second parity rule is more difficult to understand because we could not image how G and C or T and A pairs are formed in the single DNA strand. Recently, this puzzle has been solved mathematically, using the similarity of the forward and reverse strands and homogeneity of the DNA strand over the genome structure [15]. Although Chargaff's parity rules represent original intra-species phenomena, the rules can be expanded to inter-species phenomena using data from a large number of complete genomes [16]: the second parity rule is applicable only to a single DNA strand from a double-stranded DNA molecule.

Sueoka [17] was the first to analyze the cellular amino acid composition in bacteria, and our laboratory has independently analyzed the cellular amino acid compositions of bacteria, archaea, and eukaryotes [18]. Graphical representation or a diagrammatic approach to the study of complicated biological systems can provide an intuitive picture and provide useful insights [19, 20]. Using certain graphical presentations, huge data sets from genomes can be easily recognized as simple patterns representing complicated organisms. Indeed, using a radar chart to express cellular amino acid compositions, their patterns, a "star-shape," are similar among various organisms, and their differences seem to reflect biological evolution [18]. In addition, the amino acid compositions deduced from complete genomes resemble those obtained from amino acid analyses of cell lysates [21]. These results suggest that the ratios of amino acids to the total amino acids and those of nucleotides to the total nucleotide content are useful indices to characterize whole genome structures [21].

3. Patternalization of amino acid compositions

In general, there are 20 amino acids that can form proteins, and the amino acid sequences are strictly controlled by 64 codons consisting of three nucleotides, a triplet. Thus, differences in amino acid sequences of the same kind of proteins reflect biological evolution among species, although differences among different kinds of proteins seem not to be significant. Furthermore, sequence comparisons of protein mixtures are theoretically too complex to consider given currently available tools. Conversely, the amino acid composition predicted from protein(s) can characterize protein(s) from a different point of view, not only among the same organisms, but also among different organisms. In fact, the cellular amino acid compositions of various bacteria have been analyzed [17]. Based on the 20 amino acids that comprise proteins, there were 20 traits that could be evaluated, which, at first glance, seemed too many to provide meaningful information for cells.

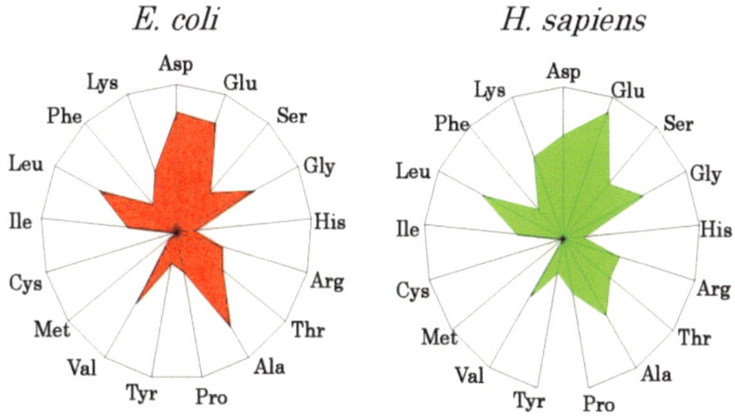

Figure 1.
Radar charts of cellular amino acid compositions of Escherichia coli *and* Homo sapiens. *Amino acid compositions are expressed as the percentage of total amino acids. Gln and Asn are combined with Glu and Asp, respectively, because the former two are converted into the latter two during hydrolysis [18].*

However, using a radar chart to present the amino acid compositions, the data could be patternalized, and the amino acid composition was observed to represent certain cellular characteristics, as shown in **Figure 1**. The patterns of bacteria (*Escherichia coli*) and of humans (*Homo sapiens*) resemble each other, although there is a great evolutionary distance between these two organisms. Microorganisms' fossils were found in 550–2800-million-year-old rocks [22–24], and it is thought that bacteria are evolutionarily close to primitive life forms. Therefore, it seemed that the primitive life forms might have similar amino acid compositions [21]. This "star-shape" cellular amino acid composition pattern must have been conserved from primitive organisms to those current organisms.

4. Chronological precedence of protein formation over codon formation

To understand the establishment of primitive organisms, the chronological precedence of protein and codon formation is a very important subject in biological evolution. Unfortunately, this theory has not yet been proven, because primitive organisms were formed under so many unknown factors an extremely long time ago. However, a simulation analysis based on a random choice of amino acids or nucleotides was carried out, which assumed that their polymerization depended on their free monomer concentrations, according to the chemical reaction rule that governs natural phenomena. Amino acid polymerizations produced a protein which reflected original free amino acid concentrations without codons, while nucleotide polymerizations did not produce functional proteins, even after considering the codon table, as shown in **Figure 2** [25]. Therefore, it seems difficult to predict "the RNA world" which presumes that RNA polymers formed primitive life forms [26]. Additionally, the possibility of the accumulation of RNA, which has a UV absorbance at around 250 nm, might be very low under the strong UV irradiation present on the primitive Earth. These results suggest that protein formation might chronologically precede codon formation at the end of prebiotic evolution, although we have no explanation of how the nucleotide sequence information necessary for proteins might have been transmitted to the nucleotide polymerization that established the codons. The

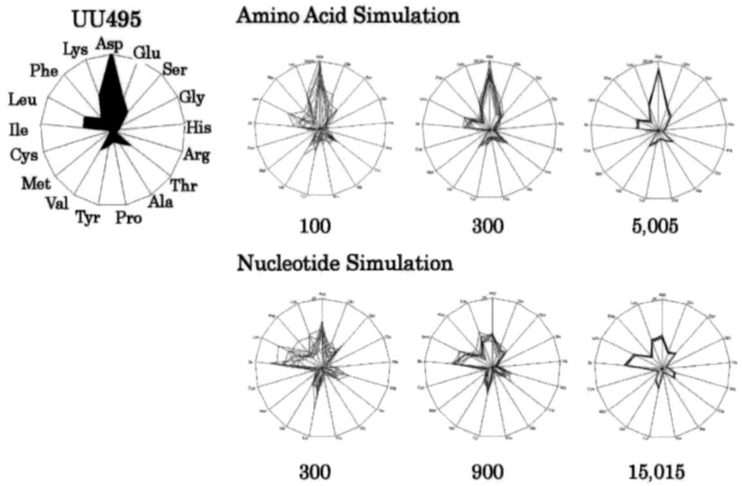

Figure 2.
Computational amino acid compositions of an Ureaplasma urealyticum *gene. Upper panel: random choice of amino acids was carried out in the original gene (5005 amino acid pool). Lower panel: random choice of nucleotides was carried out in the original gene (15,018 nucleotides). In the simulation using nucleotides, the stop codon and Trp were discarded from the calculation of amino acid compositions, and a triplet formed was immediately counted as an amino acid. This figure was adapted from Sorimachi and Okayasu [25].*

"amino acid world" [21] seems a better fit for primitive life forms rather than the "RNA world." There are several hypotheses for codon formation [27–29], but the process of codon formation has not yet been determined.

According to our simulation analyses [25], proteins that were components of primitive life forms might reflect the free amino acid concentrations on the primitive Earth. As shown in **Figure 1**, the cellular basic amino acid composition, the "star-shape," is characterized by comparatively high concentrations of hydrophobic amino acids, such as valine, leucine, and isoleucine. The glycine and alanine contents were also comparatively high. The former might contribute to self-aggregation of proteins via hydrophobicity to form primitive life forms under low protein concentrations, and the latter might reflect their easy formation on the primitive Earth. In fact, simple amino acids such as glycine and alanine have been identified in meteorites [30, 31] and can be formed by electrical discharge in an atmosphere presumed to reflect primitive Earth [32]. Conversely, the phenylalanine, tryptophan, and tyrosine content, which can absorb ultraviolet light, were quite low. Strong ultraviolet irradiation might induce photodegradation of these amino acids. The differences in amino acid contents in cellular amino acid compositions seem to reflect the presumed free amino acid concentrations on the primitive Earth and eventually resulted in the formation of the "star-shaped" cellular amino acid compositions (**Figure 1**).

5. Amino acid compositions deduced from complete genomes

Initially, amino acid compositions were deduced from complete genomes by assuming that each gene is equally expressed in a whole cell [21]. This resulted in the amino acid composition deduced from the complete genome resembling the cellular amino acid composition obtained from the amino acid analyses of cell lysates [21], as shown in **Figure 3**. This coincidence is difficult to understand because of the different origins of both values, until the genome structure has been clarified, as shown in the next section.

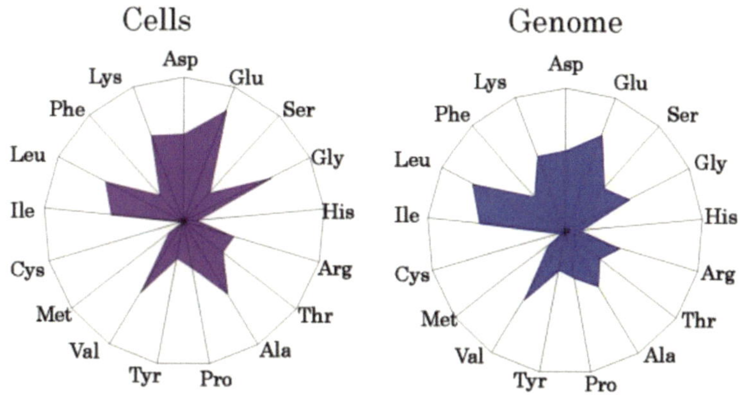

Figure 3.
Radar charts of cellular and genomic amino acid compositions. Values are expressed as the percentages of total amino acids. Pyrococcus horikoshii *was examined. The cellular amino acid composition was obtained from three independent analyses. In genomic calculations, Gln and Asn were also incorporated into Glu and Asp, respectively, to compare with data based on amino acid analysis.*

6. Homogeneity of genome structure

Each gene has its characteristic amino acid or nucleotide sequence, and its amino acid or nucleotide composition differs not only in inter-species but also in intraspecies. Conversely, gene assemblies encoding 3000–7000 amino acid

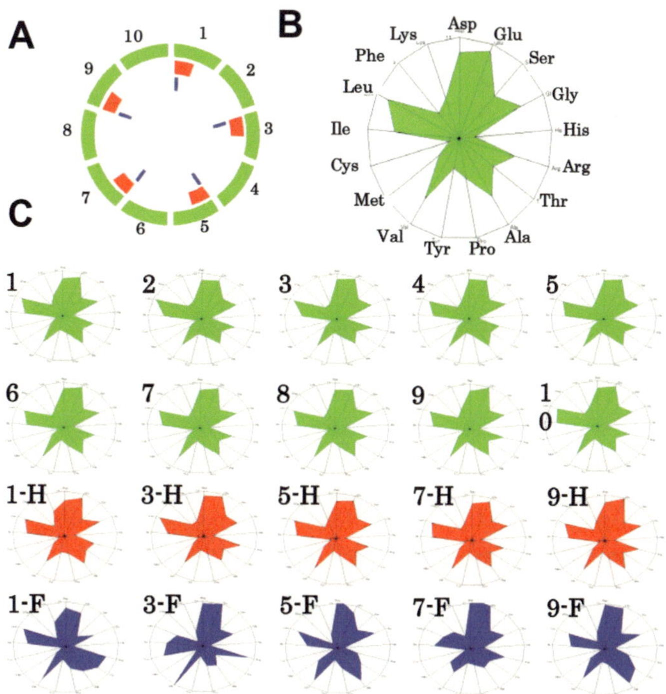

Figure 4.
Radar charts of amino acid compositions calculated from various units of the complete genome of Methanobacterium thermoautotrophicum. *(A) The complete genome structure of* M. thermoautotrophicum *(B) radar charts of amino acid compositions calculated from the complete genome, and (C) from various units. The complete genome, comprising 1869 protein genes, was divided into 10 or 20 units. Ten units (1–10); based on 186 and 195 genes, half size units (1-H–9-H); based on 93 genes, single genes (1-F–9-F); based on the first single gene of each unit. Glutamine and asparagine were calculated as glutamic acid and aspartic acid, respectively, and tryptophan (<1%) was omitted in the radar charts [18]. This figure was adapted from Sorimachi [36].*

residues show very similar amino acid compositions [33] and nucleotide compositions [34] in intraspecies examinations. Consistent results were obtained from whole chromosomes consisting of putative small units of 3000–7000 amino acid residues [33]. Additionally, it has been shown mathematically that 3000–7000 amino acid residues represent the amino acid composition of a certain amino acid pool [35]. Thus, genome structure, which is constructed homogeneously from putative similar small units, can be represented by a "pearl-necklace," as shown in **Figure 4**. The fact that the structure of a genome is homogeneously constructed with putative similar small units indicates that micro-alterations of nucleotide sequences are canceled out within the small unit and that the small unit represents the whole genome characteristics. Macro-alterations represented by the small unit, and based on species differences, occur synchronously over the genome [33]. This conclusion has never been obtained from the analysis of nucleotide or amino acid sequences of actual genes. Based on these results, the ratios of amino acids to the total amino acids or those of nucleotides to the total nucleotides form useful indices for characterizing a genome whose nucleotide numbers differ among species.

7. Nucleotide compositions

As described above, the intraspecies rule of nucleotide composition was reported by Chargaff in 1950, as the first parity rule [12], and a similar parity rule regarding the single DNA strand was reported by the same group in 1968, as the second parity rule [14]. Using the normalized values to 1 ($G + C + T + A = 1$), the following relationships are obtained: $G = C$, $T = A$, and $[(G + A) = (C + T)]$. Recently, Mitchell and Bridge [16] reported that Chargaff's second parity rule is applicable to a single DNA strand comprising a double-stranded DNA, based on many complete genome data among various species. Conversely, we showed that chloroplast and plant mitochondrial DNA and nuclear DNA obey Chargaff's second parity rule as an inter-species rule [37], and that the second parity rule was applicable to the nucleotide relationships not only in the coding region, but also in non-coding regions compared with those of the complete single DNA strand [37, 38]. When invertebrate mitochondrial DNA is classified into two groups, high C/G and low C/G ratios, nucleotide content relationships may be expressed by linear formulae [37]. However, organellar DNA deviated from Chargaff's second parity rule and nucleotide relationships were heteroskedastic [16, 39, 40]. The fact that all regression lines based on different kingdoms closed at the same single point suggests that all species descended from a single origin [41]. This is the first demonstration based on scientific evidence that all species were descended from a single origin of life. This concept has been presumed since Darwin's theory "Origin of Species" was published in 1859. Charles Darwin discussed evolution over the course of generations via the presence of "Natural Selection" in "On the Origin of Species by Means of Natural Selection, or the Preservation of Favoured Races in the Struggle for Life"; however, he discussed neither "a single origin" nor "a common ancestor" of species. The two regression lines of nucleotide relationships based on coding and non-coding regions closed to form a wedge-shape, because both fragments exist on the same DNA strand [37]. Similarly, the two regression lines based on chloroplast and plant mitochondrial DNA also closed to form a wedge-shape [37]. Thus, both organellar DNA independently descended from the same origin in biological evolution. Quite recently, it has been shown that vertebrates are descended from a certain invertebrate [42]. However, although the phylogenetic trees [7–11] have an apparent single origin, these "facts" are merely mathematical calculation results.

8. Diagonal genome universe

Chargaff's parity rules were originally based on intraspecies phenomena [12, 14], and the rules are applicable to inter-species evolutionary phenomena for nuclear, chloroplast, and plant mitochondria as mentioned above. The rules are represented by the following equations: G = C, T = A, [(G + A) = (C + T)]. As all values are normalized to 1, Chargaff's parity rule can also be represented as: 2G + 2A = 1, A = 0.5 – G, T = 0.5 – G, C = G, G = (G). The lines G and C overlap and the lines A and T overlap, and the former is line symmetrical to the latter against the line y = 0.25, as shown in **Figure 5**. These equations mean that four nucleotide contents can be expressed by just one nucleotide content using regression lines (**Figure 5**), and the two duplicate nucleotide contents (G or C and T or A) are symmetrical. Thus, the four nucleotide contents (two duplicate points) move strictly on the diagonal of 0.5 of a square in nuclear, chloroplast, and mitochondrial DNA, which obey Chargaff's second parity rule. Therefore, biological evolution caused by nucleotide alterations is expressed on the diagonal of a 0.5 square: the "diagonal genome universe" [36], although biological evolution shows a wide spectrum of phenotypic expressions over a 3.5-billion-year period.

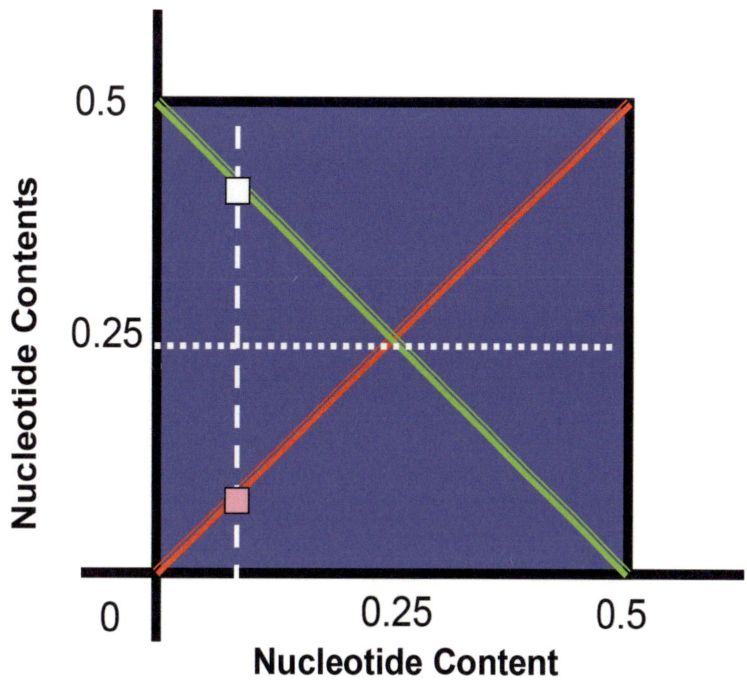

Figure 5.
The "Diagonal Genome Universe." Plotting four nucleotide contents normalized to 1 against certain nucleotide content (i.e., G or C content), G and C contents are expressed by (G = G) and (G = C), respectively, and T and A contents are expressed by (T = 0.5 – G) and (A = 0.5 – G), respectively. For example, if G = 0.1 (white dashed line), C = 0.1, T = 0.4, and A = 0.4. White open square, A or T; pink closed square, C or G. The white dotted line represents the line of symmetry (y = 0.25). Similarly, plotting nucleotide contents against T or A content, (T = T), (T = A), (C = 0.5 – T or A), and (G = 0.5 – T or A) are obtained. This figure was adapted from Sorimachi [36].

9. Codon evolution

The 20 amino acids are encoded by genes using nucleotide triplets; therefore, these sequences are determined according to triplet sequences. Additionally, amino acid sequences differ not only inter-gene but also intraspecies. These facts indicate that a comparison of codon evolution based on the complete genome, which comprises large numbers of different genes, would not be significant. Indeed, no clear evaluation has been obtained, despite the attempted explanations of many scientists [27–29]. However, as described in the previous section, it has been clarified that a whole genome is constructed from putative small units that encode proteins of similar amino acid composition. This suggests that the total codon usage deduced from the complete genome is stable and represents the whole genome characteristic. According to this concept, correlationships of nucleotide contents in a complete genome can be expressed by the linear formula, y = ax + b; where "y" and "x" are nucleotide contents, and "a" and "b" are constant values. In addition, as each codon usage is expressed by a linear formula among various organisms, the determination of any one nucleotide content in certain organism can essentially estimate other three nucleotide contents and, therefore, the 64 codon usages (**Figure 6**). The estimated codon usage patterns and amino acid compositions are almost the same between the original experimental results and estimated results. The codon usage patterns clearly indicate that codon usages changed synchronously among the 64 codons during biological evolution.

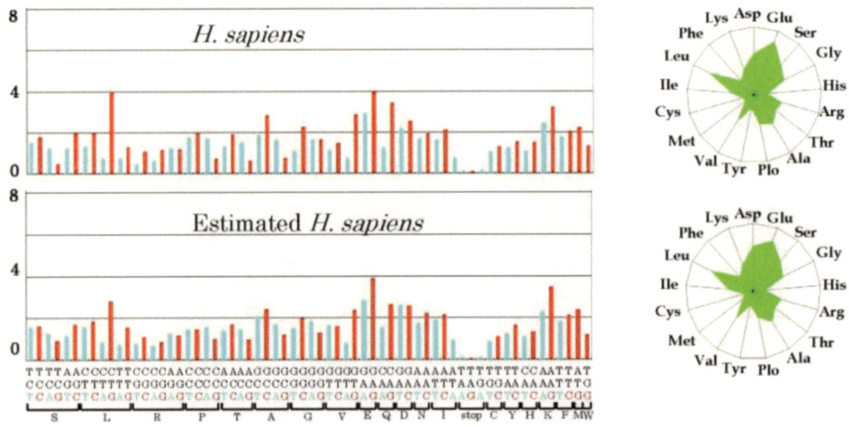

Figure 6.
Codon usage patterns and amino acid compositions of Homo sapience. *Codon usage (bar) and amino acid composition (radar chart) are expressed as a percent of total codons and amino acids, respectively. Upper and lower panels represent genomic and estimated data, respectively. This figure was reproduced from Sorimachi and Okayasu [38].*

10. Natural selection in biological evolution based on amino acid contents

The above mentioned theories have been described in previous review articles [36, 43]; therefore, in this section, unique applications based on the amino acid compositions or nucleotide contents in the construction of phylogenetic trees to study evolution are presented using recent data.

The theory of natural selection was promoted by Charles Darwin and Alfred Wallace 150 years ago. This theory was derived from specific differences or similarities in the phenotypes of organisms that lived on geologically isolated islands.

DOI: http://dx.doi.org/10.5772/intechopen.91170

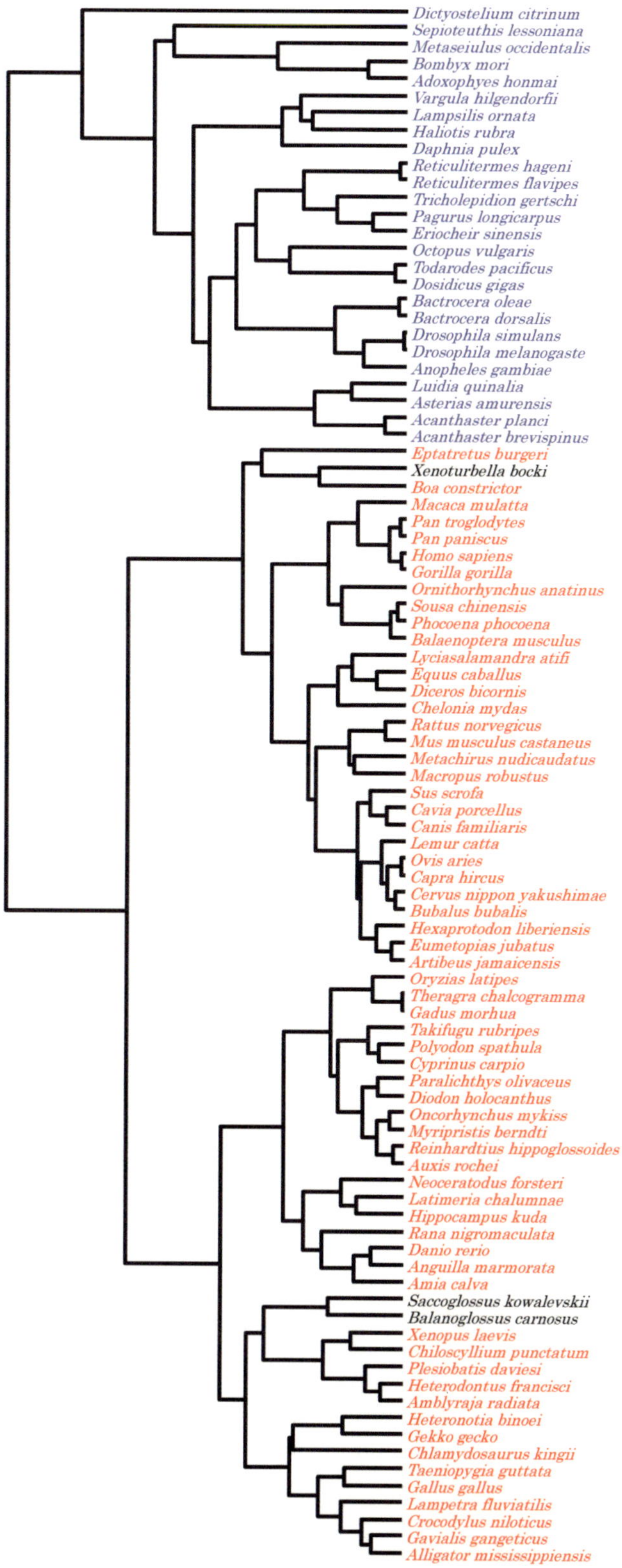

Figure 7.
Phylogenetic tree generated using Ward's cluster analysis method [48] from the predicted amino acid composition of the complete mitochondrial genomes of 26 invertebrates (blue), 3 hemichordates (black), and 63 vertebrates (red). This figure first appeared in Ref. [49] and is reproduced with permission.

The theory of biological evolution has been further developed by paleontology [44], using phenotypic changes in fossils, and by molecular biology [6], using genotypic modifications (nucleotides or amino acids) of genes in living organisms.

Generally, the nucleotide or amino acid sequences of a particular gene or genes have been the focus of biological evolution studies, and many phylogenetic trees have been constructed using nucleotide or amino acid sequences [7–11, 27, 29, 45]. Conversely, the amino acid compositions or nucleotide contents have been rarely used for whole genome research. However, these indices have been used to classify bacteria, archaea, and eukaryotes [46] and recently vertebrate evolution [47]. In those studies, all organisms could be classified into two types, "GC-rich" and "AT-rich," and the vertebrates examined were further classified into two groups: terrestrial and aquatic vertebrates, based on natural selection. A similar result was obtained from an analysis based on 16S rRNA sequences [45, 47].

When the normalized amino acid compositions of vertebrate and invertebrate complete mitochondrial genomes were used, the groups were separated cleanly into two large clusters, vertebrates and invertebrates (**Figure 7**). In invertebrates, starfish (Echinodermata) formed a small cluster, and squids and octopus (Mollusca) were grouped into the same cluster. Vertebrates were further classified into three major clusters, mammals, fish, and a mixture of reptiles and amphibians. For example, primates (human, chimpanzee, and gorilla) formed a small cluster. Thus,

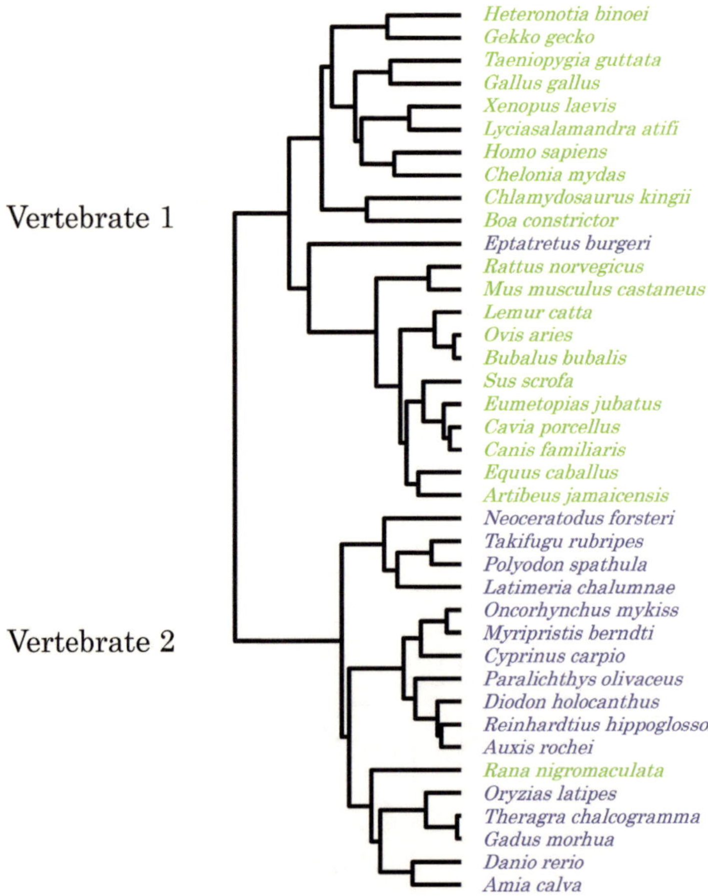

Figure 8.
Phylogenetic tree of complete vertebrate mitochondrial genomes based on cluster analysis [51] using amino acid compositions as the trait. Green and blue characters represent terrestrial and aquatic vertebrates, respectively. This figure was adapted from Sorimachi et al. [47].

close species fell into the same cluster and did not split into different clusters. These results indicate that the normalized values of amino acid and nucleotide contents calculated from complete genomes could be used to characterize organisms and to construct phylogenetic trees. Our results based on complete mitochondrial genomes revealed that hemichordates (*Balanoglossus carnosus* and *Saccoglossus kowalevskii*) and *Xenoturbella bocki*, which were classified into the low G/C content invertebrates group, were closer to vertebrates than to invertebrates [49]. Protists (*Monosiga brevicollis*) and cephalochordate (*Branchiostoma belcheri*) were classified into the low G/C and high G/C content invertebrate groups, respectively [49].

In a previous study to classify vertebrates [49, 50], as organisms were chosen at random without any preposition, it was difficult to evaluate whether the classification results were reasonable in the phylogenetic trees. Using the amino acid composition as the trait, the vertebrates examined were separated into two major clusters (**Figure 8**), terrestrial and aquatic vertebrates. The exceptions were the hagfish (*Eptatretus burgeri*), which fell into the terrestrial vertebrate cluster, and the black spotted frog (*Rana nigromaculata*), which clustered with the aquatic vertebrates [47]. The clustering of the

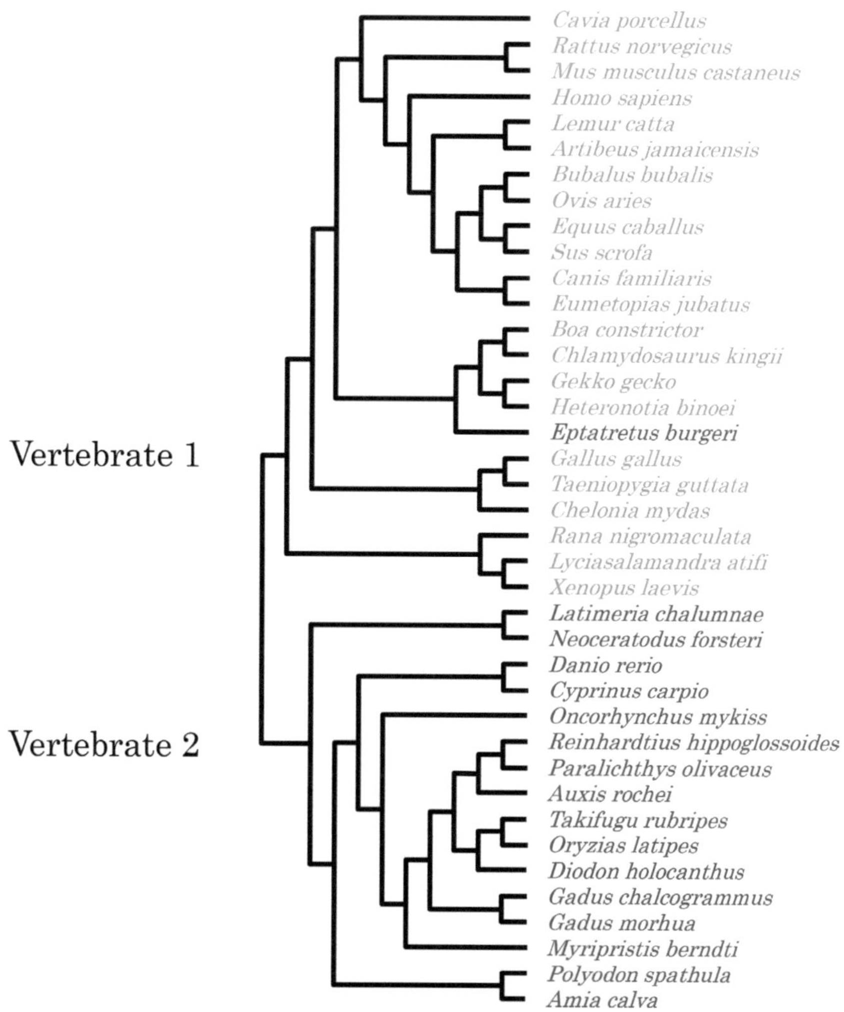

Figure 9.
Phylogenetic tree of 16S rRNA. The phylogenetic tree was constructed by the neighbor-joining method [48] using nucleotide sequences. Green and blue characters represent terrestrial and aquatic vertebrates, respectively. This figure was adapted from Sorimachi et al. [47].

hagfish (*E. burgeri*) with the terrestrial vertebrates may reflect the controversy over the classification of this fish [52]. If the hagfish truly belongs to the terrestrial group, it suggests that hagfish still possesses some primitive mitochondrial characteristics that were present before its evolution. The frog (*R. nigromaculata*) was consistently grouped with the aquatic vertebrates which may reflect the conservation of tadpole characteristics after metamorphosis. The coelacanth (*Latimeria chalumnae*), the Queensland lungfish (*Neoceratodus forsteri*), which is a living fossil and one of the oldest living vertebrate genera, and the American paddlefish (*Polyodon spathula*), which is the oldest living animal species in North America, all belonged to an additional small cluster. Using the G, C, A, and T content of the coding regions, non-coding regions, and complete mitochondrial genomes as the traits in cluster analyses, similar results were obtained, but with some additional exceptions [50].

Single genes have been used to construct phylogenetic trees [7–11], and 16S rRNA has been frequently examined [27, 29]. The phylogenetic tree based on 16S rRNA sequences of various vertebrates is shown in **Figure 9**. The tree is consistent with that based on nucleotide contents. The hagfish (*E. burgeri*) fell into the terrestrial vertebrates, while the black spotted frog (*R. nigromaculata*) belonged to the terrestrial vertebrates. These results indicate that vertebrate evolution is controlled by natural selection under both an internal bias resulting nucleotide replacement rules and by an external bias caused by environmental biospheric conditions. In addition, based on amino acid composition or nucleotide content of complete mitochondrial genomes, Hemichordates (*Balanoglossus carnosus* and *Saccoglossus kowalevskii*) and Xenoturbella were classified into vertebrates not into invertebrates [49].

11. Organelle evolution

In Chargaff's first parity rule [12], G = C and A = T in a double DNA strand, while in the second parity rule [14], G ≈ C and A ≈ T in a complete single DNA strand. Based on Chargaff's second parity rule, nucleotide content differences such as (G – C) and (A – T) reflect biological evolution. In addition, the other nucleotide content differences, (G – A, G – T, C – A, and C – T), also reflect biological evolution [34, 53].

Six nucleotide content differences among the complete mitochondria of the four species (*M. brevicollis, P. pallidum, D. discoideum*, and *R. Americana*) were examined (**Figure 10**, left panel). The GC and AT skew are expressed by the ratios of (G – C)/(G + C) and (A – T)/(A + T), respectively [54]. The skew seems to be due to differences in replication processes between the leading and lagging strands [55]. In the replication of the lagging strand, the deamination of cytosine increases the probability of mutations, and the inversion of nucleotide content differences reflects biological divergence. Similarly, these phenomena are observed in mitochondria, consisting of heavy (H) and light (L) chains [56–58]. When the GC skew was plotted against G content, animal mitochondria were classified into two groups: high and low C/G [59].

To allow simple comparison of inter- and intraspecies genome structures, genomes were divided into three fragments throughout subsequent analyses, from which three separate patterns emerged. There is no inversion of nucleotide content differences that was observed in the mtDNA of *M. brevicollis* (G: 0.081, C: 0.059), the mycetozoan *Polysphondylium pallidum* (G: 0.143, C: 0.085), or *Dictyostelium discoideum* (G: 0.171, C: 0.104) (**Figure 10**), whereas differences in (G – C) and (T – A) values for *M. brevicollis* mtDNA were the lowest among these species. Choanoflagellates are most closely related to animals based on genome sequencing [60]. The fact that the nucleotide content difference patterns of the three fragments were almost identical for these three species indicates that their nucleotide distributions were homogeneous, and that the nucleotide content was symmetrical.

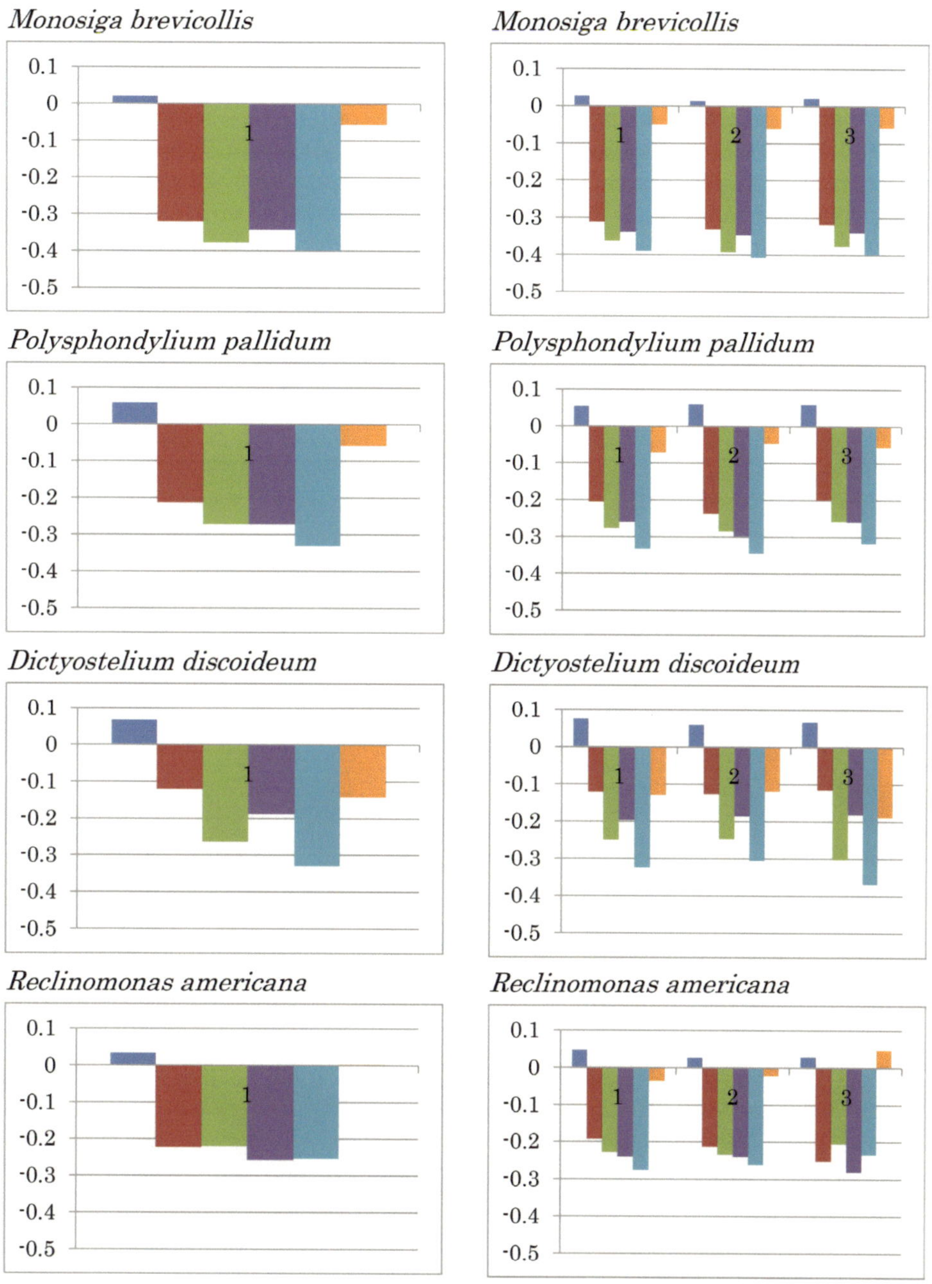

Figure 10.
Nucleotide content differences in complete mitochondrial genomes (left side) and the three fragments of each mitochondrial genome (right side). Left to right: (G – C), (G – T), (G – A), (C – T), (C – A), and (T – A).

Based on these results, these mitochondria are likely to be primitive. Consistent results were obtained from Ward's clustering analysis using amino acid compositions predicted from complete mitochondrial genomes as traits [59]. Thus, the *M. brevicollis* mitochondrion is the most primitive among the three. Although the *Reclinomonas americana* mtDNA (G: 0.148, C: 0.114) has previously been proposed as a mitochondrial ancestor [61], AT inversion was observed in the third fragment. In addition, differences in (G – C) and (T – A) values in *R. americana* mtDNA were smaller than those in the mtDNA of the previous three organisms. The unsymmetrical nucleotide content causes significant differences in nucleotide content patterns as a result of nucleotide content inversion. Judging from these results, the *R. americana* mitochondrion is probably more evolved than the former three mitochondria. In addition, AT inversion occurred in the following more highly evolved organisms: Mollusca species, squid (*Todarodes pacificus*), octopus (*Octopus vulgaris*), Echinodermata species, sea urchin (*Paracentrotus lividus*), water flea (*Daphnia pulex*), hermit crab (*Pagurus longicarpus*), and Humboldt squid (*Dosidicus gigas*) [53, 62]. In addition, large positive (G – A) values in the three fragments were observed in *Paragonimus westermani*, while large positive (G – C) and (A – T) values in the three fragments were observed for the mtDNA of representatives of the following phyla: Cnidaria (*Pavona clavus*), Platyhelminthes (*Schistosoma mansoni*), Porifera (*Geodia neptuni*), Arthropoda (*Tigriopus californicus*), and Chordata (*Branchiostoma belcheri*) [53]. Furthermore, the following invertebrate

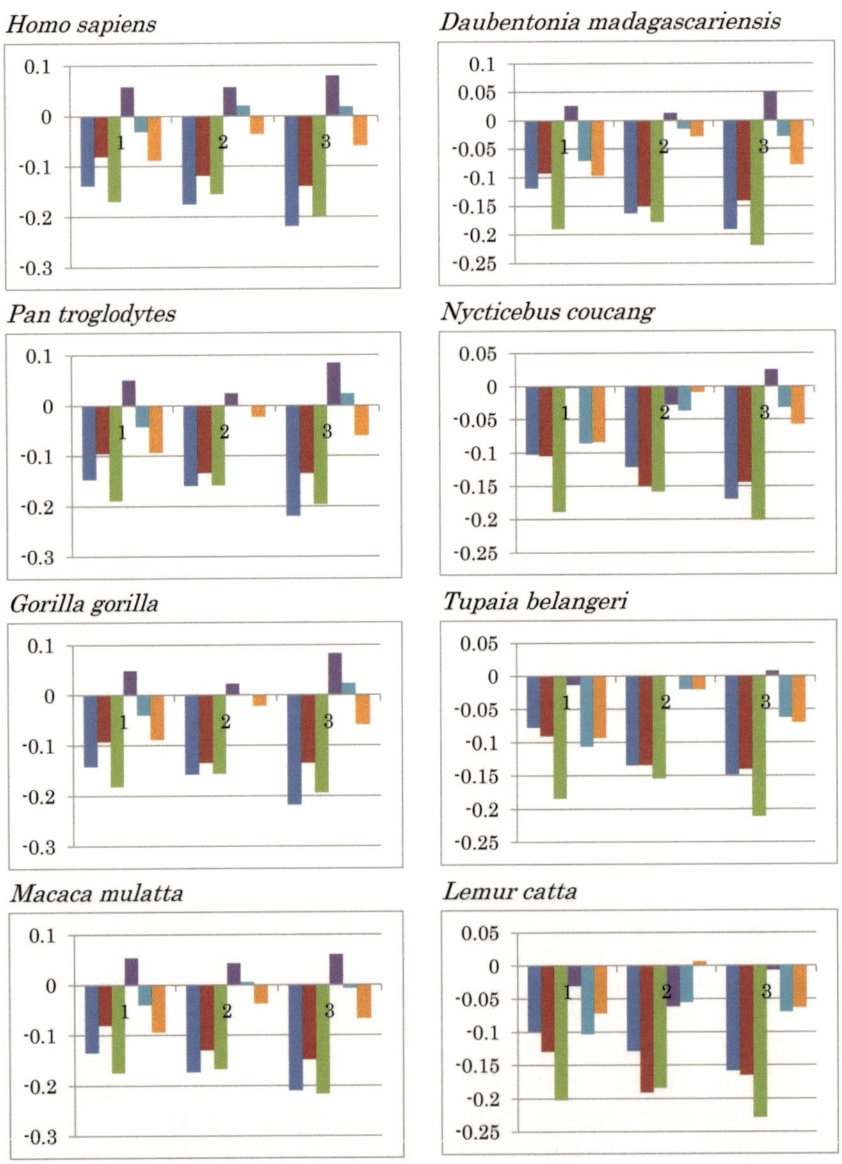

Figure 11.
Nucleotide differences in the three fragments of each primate mitochondrial genome. Left to right: (G – C), (G – T), (G – A), (C – T), (C – A), and (T – A).

mitochondria were also examined: *Acanthaster planci*, *Haliotis rubra*, *Lampsilis ornate,* and the mtDNA of hemichordates, *Saccoglossus kowalevskii, Balanoglossus carnosus*, and *Xenoturbella bocki* was examined [53].

In the mtDNA of primate species *H. sapiens*, *P. troglodytes*, *G. gorilla*, *Macaca mulatta*, *Daubentonia madagascariensis*, *Nycticebus coucang*, and *Tupaia belangeri*, nucleotide content difference patterns were quite similar in the first four species, and large positive increases in (C – T) differences in the three fragments clearly indicated evolutionary divergence (**Figure 11**). The positive (C – T) differences in all three fragments were characteristic of these four primate mitochondria, while positive increases in (C – T) values were only observed in the third fragment of *N. coucang* and *T. belangeri* mtDNA. In contrast, nucleotide content difference patterns of the prosimian *Lemur catta* completely differed from those of the primates, although TA inversion was observed in the second fragment. The primate mtDNA nucleotide content patterns were also completely different from that of hemichordate *B. carnosus*, although their C contents were the highest among all organisms examined [59]. This finding indicates that mitochondrial structures respect epigenomic evolutionary functions.

12. Definitive universal equations

In the normalization of nucleotide contents (G + C + A + T = 1), as (G = C) and (A = T) based on Chargaff's parity rules, (2G + 2A = 1) is obtained. This equation is altered to (A = 0.5 – G) and then (A – G = 0.5 – 2G). Finally, G – A = 2G – 0.5. The relationship between (G – A) and (G) is linear when both (G) and (A) are expressed by linear functions. In animal mitochondria, only the correlations between the two purines (A versus G) or the two pyrimidines (C versus T) are linear, while the correlations between purines and pyrimidines (A or G versus T or C) are weak or not correlated at all [62]. For example, when plotting (G – C), (G – T), (G – A), (C – T), (C – A), and (T – C) against G content, only (G – A) versus G content was linear in vertebrate mitochondria [59]. In invertebrate mitochondria, plotting nucleotide content differences against G content was weakly linear.

Plotting (X – Y)/(X + Y) against (X – Y), the following linear relationship was obtained in mitochondria, chloroplasts, and chromosomes (**Figure 12**): (X – Y)/(X + Y) = a (X – Y) + b, where X and Y are nucleotide contents, and (a) and (b) are constants. As (b) was almost null and (a) was ~2.0, (X – Y)/(X + Y) ≈ 2.0 (X – Y). In these genome analyses, which are independent of Chargaff's parity rules, the values of (a) for (G, C), (G, A), (G, T), (C, T), (C, A), and (A, T) were 2.5858, 1.85558, 1.9908, 1.9771, 1.9968, and 1.5689, respectively, in our previous results [53, 54]. Based on these results, (G + C), (G + A), (G + T), (C + A), (C + T), and (A + T) were 0.39, 0.54, 0.50, 0.51, 0.50, and 0.64, respectively. In virus genome analyses [53, 54], the constant values for (a) were 1.9–2.1, and the values for (X + Y) were 0.47–0.53. In contrast, in the normalization of nucleotide contents (G + C + A + T = 1), as (G = C) and (A = T) based on Chargaff's parity rules, (2G + 2A = 1) is obtained. This equation is altered to (G + A = 0.5). This value is consistent with the value obtained above from genome analyses. Similarly, (G + T = 0.5), (C + A = 0.5), and (C + T = 0.5), although (G + C) and (A + T) cannot be determined. Therefore, the four nucleotide contents are expressed by the following regression lines, plotted against G content: A = 0.5 – G, T = 0.5 – G, C = G, and G = G. Lines G and C overlap, as do lines A and T, and the former line is symmetrical to the latter against line (y = 0.25). The intercepts of lines G and C are close to the origin, while those of lines A and T are close to 0.5 at the vertical and horizontal axes. All organisms from bacteria to *H. sapiens* are located on the diagonal lines of a 0.5 square, termed the "Diagonal Genome Universe," using the normalized values that obey Chargaff's first parity rule [12]. These relationships lead to (G or C) + (A or T) = 0.5. The present results indicate that a linear regression line equation, (X – Y)/(X + Y) = a (X – Y) + b, universally represents all normalized values, including the values deviating from Chargaff's parity rules. This newly discovered equation clearly reflects not only Chargaff's first parity rules, based on hydrogen bonding between two nucleotides, but also natural rule.

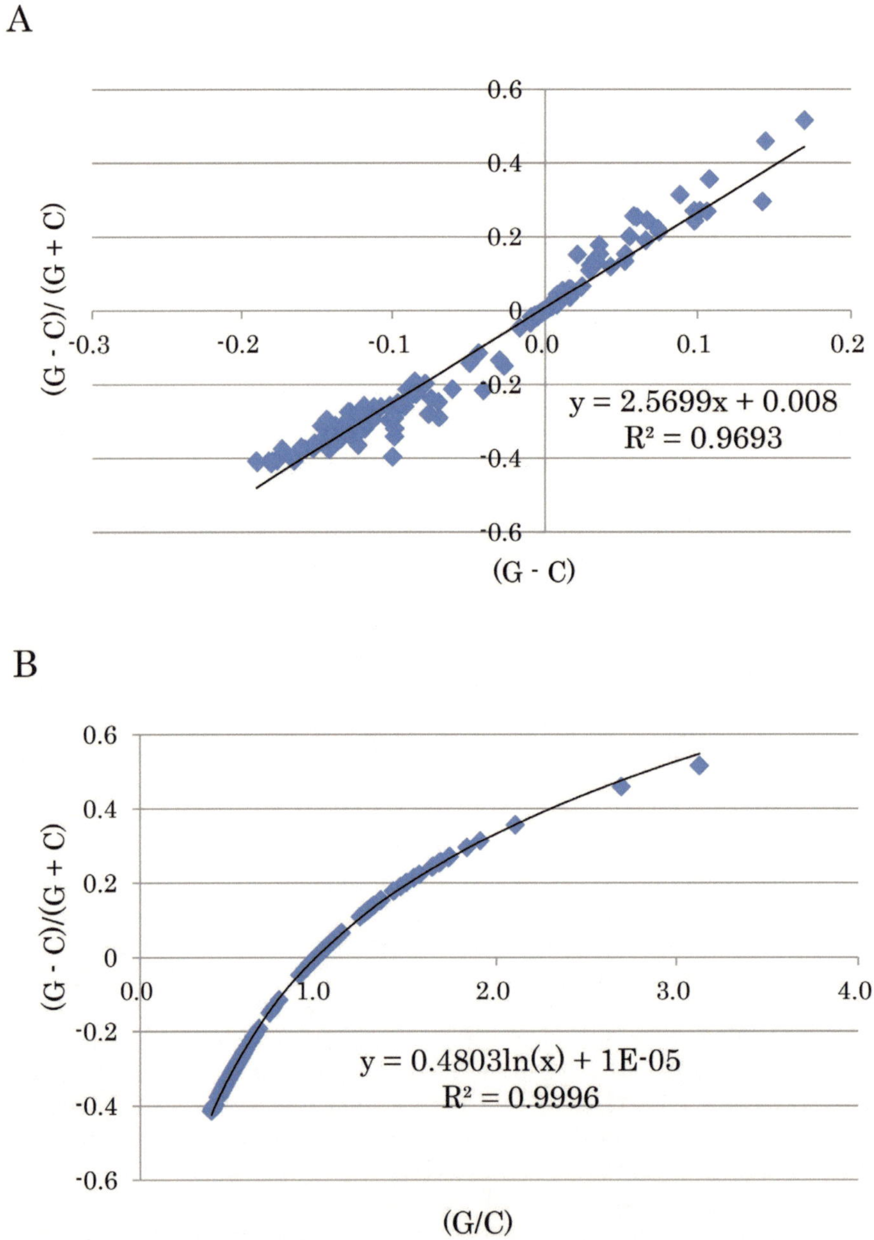

Figure 12.
Universal rules. The following genome samples were examined: mitochondria of vertebrates (65), invertebrates (54), and non-animals (42), chloroplasts (28), prokaryote chromosomes (21), and eukaryote chromosomes (15). Left side: relationship between (X – Y) and (X – Y)/(X + Y) and right side: relationship between (X/Y) and (X – Y)/(X + Y).

A linear regression line was not obtained when using randomly chosen value (**Figure 12A**). Furthermore, plotting (X – Y)/(X + Y) against (X/Y), the following logarithmic function was obtained for all tested genomes as well as when using randomly chosen values (**Figure 12B**): (X – Y)/(X + Y) = a ln (X/Y) + b. As (b) was almost null and (a) was ~0.5, (X – Y)/(X + Y) ≈ 0.5 ln (X/Y). The ratio between two values, (X/Y), can be expressed by a logarithmic function, ~0.5 ln (X/Y) ≈ (X – Y)/(X + Y). Plotting the GC skew vs. G content, animal mitochondria were classified into two groups: high and low C/G [59]. This fact indicates that the ratio C/G and the GC skew are evolutionarily related to each other. Any change can be expressed universally by a definitive logarithmic function, (X – Y)/(X + Y) = a ln (X/Y) + b. The present results indicate that cellular organelle evolution is strictly controlled under these characteristic rules, although non-animal mitochondria, chloroplasts, and chromosomes are controlled under Chargaff's parity rules [12, 14]. The present study clearly shows that biological evolution, which seems to be based on complicated processes, is governed by simple universal equations.

13. Conclusions

The ratios of amino acids to the total amino acids or of nucleotides to total nucleotides predicted from complete genomes consisting of huge number of nucleotides can characterize a whole organism. In addition, as these values are independent of species and genome size, these indexes are very useful for genome research, as well as single gene research. The validity of these indexes is clearly based on the homogeneity of genomic structures. In addition, patternalization of values after simple calculations based on large data sets can provide an intuitive picture and provide useful insights, revealing the homogeneity of genomic structures followed by synchronous alterations over the genome. In addition, any change between two values, X and Y, including biological evolution can be expressed definitively by a linear regression line equation, (X – Y)/(X + Y) = a (X – Y) + b, where X and Y are nucleotide contents, and (a) and (b) are constants, and by a logarithmic function, (X – Y)/(X + Y) = a′ ln (X/Y) + b′, where (a′) and (b′) are constants. As the present review is based on the endeavors and data of numerous scientists from all over the world, the author would like to express finally his following feeling as one of scientists. (Human being is an organism of huge numbers of organisms on the Earth, and we are not ranked as a special species above all organisms as a result of long evolution.) However, we have made the present modern civilization based on fossil energy usage which seems to induce climate changes. Thus, we must be responsible to establish sustainable development not only for Human being but also for other organisms. The Earth is for all organisms, not only for Human being.

Acknowledgements

The author greatly acknowledges President Hiroyuki Okada of Shinko Sangyo, Co. Ltd., Takasaki, Gunma, Japan for his financial support and Dr. Teiji Okayasu who was one of collaborators in Dokkyo Medical University for his excellent computer analyses.

Author details

Kenji Sorimachi[1,2]

1 Educational Support Center, Dokkyo Medical University, Tochigi, Japan

2 Research Laboratories, Gunma Agriculture and Forest Development Com., Ltd., Takasaki, Gunma, Japan

*Address all correspondence to: kenjis@jcom.home.ne.jp

References

[1] Sanger F, Coulson AR. A rapid method for determining sequences in DNA by printed synthesis with DNA polymerase. Journal of Molecular Biology. 1975;**94**:441-446

[2] Maxam AM, Gilbert W. A new method for sequencing DNA. Proceedings of the National Academy of Sciences of the United States of America. 1977;**74**:560-564

[3] Fleischmann RD, Adams MD, White O, Clayton RA, Kirkness EF, Kerlavage AR, et al. Whole-genome random sequencing and assembly of *Haemophilus influenzae* Rd. Science. 1995;**269**:496-512

[4] Lander ES, Linton ML, Birren B, Nusbaum C, Zody MC, Baldwin J, et al. Initial sequencing and analysis of the human genome. Nature. 2001;**409**:860-921

[5] Venter JC, Adams MD, Myers EW, Li PW, Mural RJ, Sutton GG, et al. The sequence of the human genome. Science. 2001;**291**:1304-1351

[6] Zuckerkandl E, Pauling LB. Molecular disease, evolution, and genetic heterogeneity. In: Kasha M, Pullman B, editors. Horizons in Biochemistry. New York: Academic Press; 1962. pp. 189-225

[7] Dayhoff MO, Park CM, McLaughlin PJ. Building a phylogenetic trees: Cytochrome C. In: Dayhoff MO, editor. Atlas of Protein Sequence and Structure. Vol. 5. Washington, D.C.: National Biomedical Foundation; 1977. pp. 7-16

[8] Sogin ML, Elwood HJ, Gunderson JH. Evolutionary diversity of eukaryotic small subunit rRNA genes. Proceedings of the National Academy of Sciences of the United States of America. 1986;**83**:1383-1387

[9] Doolittle WF, Brown JR. Tempo, mode, the progenote, and the universal root. Proceedings of the National Academy of Sciences of the United States of America. 1994;**91**:6721-6728

[10] Maizels N, Weiner AM. Phylogeny from function: Evidence from the molecular fossil record that tRNA originated in replication, not translation. Proceedings of the National Academy of Sciences of the United States of America. 1994;**91**:6729-6734

[11] DePouplana L, Turner RJ, Steer BA, Schimmel P. Genetic code origins: tRNAs older than their synthetases? Proceedings of the National Academy of Sciences of the United States of America. 1998;**95**:11295-11300

[12] Chargaff E. Chemical specificity of nucleic acids and mechanism of their enzymatic degradation. Experientia. 1950;**VI**:201-209

[13] Watson JD, Crick FHC. Genetical implications of the structure of deoxyribonucleic acid. Nature. 1953;**171**:964-967

[14] Rundner R, Karkas JD, Chargaff E. Separation of *B. subtilis* DNA into complementary strands. 3. Direct analysis. Proceedings of the National Academy of Sciences of the United States of America. 1968;**60**:921-922

[15] Sorimachi K. A proposed solution to the historic puzzle of Chargaff's second parity rule. Open Genomics Journal. 2009;**2**:12-14

[16] Mitchell D, Bridge R. A test of Chargaff's second rule. Biochemical and Biophysical Research Communications. 2006;**340**:90-94

[17] Sueoka N. Correlation between base composition of deoxyribonucleic acid

and amino acid composition in proteins. Proceedings of the National Academy of Sciences of the United States of America. 1961;**47**:1141-1149

[18] Sorimachi K. Evolutionary changes reflected by the cellular amino acid composition. Amino Acids. 1999;**17**:207-226

[19] Chou K-C. Review: Applications of graph theory to enzyme kinetics and protein folding kinetics. Steady and non-steady state systems. Biophysical Chemistry. 1990;**35**:1-24

[20] Qi XQ, Wen J, Qi ZH. New 3D graphical representation of DNA sequence based on dual nucleotides. Journal of Theoretical Biology. 2007;**249**:681-690

[21] Sorimachi K, Itoh T, Kawarabayasi Y, Okayasu T, Akimoto K, Niwa A. Conservation of the basic pattern of cellular amino acid composition during biological evolution and the putative amino acid composition of primitive life forms. Amino Acids. 2001;**21**:393-399

[22] Schopf JW, Barghoorn ES, Maser MD, Gordon RO. Electron microscopy of fossil bacteria two billion years old. Science. 1965;**149**:1365-1367

[23] MacGregor IM, Truswell JF, Eriksson KA. Filamentous alga from the 2,300 m.y. old Transvaal dolomite. Nature. 1974;**247**:538-539

[24] Nagy LA, Zumberge JE. Fossil microorganisms from the approximately 2800 to 2500 million-year-old Bulawayan stromatolite: Application of ultramicrochemical analyses. Proceedings of the National Academy of Sciences of the United States of America. 1976;**73**:2973-2976

[25] Sorimachi K, Okayasu T. Mathematical proof of the chronological precedence of protein formation over codon formation. Current Topics in Peptide & Protein Research. 2007;**8**:25-34

[26] Gilbert W. The RNA world. Nature. 1986;**319**:618

[27] Crick FHC. The origin of genetic code. Journal of Molecular Biology. 1968;**38**:367-379

[28] Wong JT-F. A co-evolutionary theory of the genetic code. Proceedings of the National Academy of Sciences of the United States of America. 1975;**72**:1909-1912

[29] Woese CR. Order in the genetic code. Proceedings of the National Academy of Sciences of the United States of America. 1965;**54**:71-75

[30] Kvenvolden K, Lawless J, Pering K, Peterson E, Flores J, Ponnamperuma C, et al. Evidence for extraterrestrial amino-acids and hydrocarbons in the Murchison meteorite. Nature. 1970;**228**:923-926

[31] Wolman Y, Haverland W, Miller SL. Nonprotein amino acids from spark discharges and their comparison with the Muchison meteorite amino acids. Proceedings of the National Academy of Sciences of the United States of America. 1972;**69**:809-811

[32] Miller SL. A production of amino acids under possible primitive earth conditions. Science. 1953;**117**:528-529

[33] Sorimachi K, Okayasu T. Gene assembly consisting of small units with similar amino acid composition in the *Saccharomyces cerevisiae* genome. Mycoscience. 2003;**44**:415-417

[34] Sorimachi K, Okayasu T. An evolutionary theories based on genomic structures in *Saccharomyces cerevisiae* and *Enchephalitozoon cuniculi*. Mycoscience. 2004;**45**:345-350

[35] Sorimachi K, Okayasu T, Ebara Y, Nakagawa T. Mathematical proof of genomic amino acid composition homogeneity based on putative small units. Dokkyo Journal of Medical Sciences. 2005;**32**:99-100

[36] Sorimachi K. Evolution based on genome structure: The "diagonal genome universe". Natural Science. 2010;**2**:1104-1112

[37] Sorimachi K, Okayasu T. Universal rules governing genome evolution expressed by linear formulas. Open Genomics Journal. 2008;**1**:33-43

[38] Sorimachi K, Okayasu T. Codon evolution is governed by linear formulas. Amino Acids. 2008;**34**:661-668

[39] Bell SJ, Forsdyke DR. Deviations from Chargaff's second parity rule with direction of transcription. Journal of Theoretical Biology. 1999;**197**:63-76

[40] Nikolaou C, Almirantis Y. Deviations from Chargaff's second parity rule in organelle DNA insights into the evolution of organelle genomes. Gene. 2006;**381**:34-41

[41] Sorimachi K. Genomic data provides simple evidence for a single origin of life. Natural Science. 2010;**2**:519-525

[42] Sorimachi K, Okayasu T, Ohhira S, Fukasawa I, Masawa N. Evidence for the independent divergence of vertebrate and high C/G ratio invertebrate mitochondria from the same origin. Natural Science. 2012;**4**:479-483

[43] Sorimachi K. Evolution from primitive life to *Homo sapiens* based on visible genome structures: The amino acid world. Natural Science. 2009;**1**:107-119

[44] Cobbett A, Wilkinson M, Wills M. Fossils impact as hard as living taxa in parsimony analyses of morphology. Systems Biology. 2007;**17**(2007):753-766

[45] Weisburg WG, Barns SM, Pelletier DA, Lane DJ. 16S ribosomal DNA amplification for phylogenetic study. Journal of Bacteriology. 1991;**173**:697-703

[46] Okayasu T, Sorimachi K. Organisms can essentially be classified according to two codon patterns. Amino Acids. 2009;**36**:261-271

[47] Sorimachi K, Okayasu T, Ohhira S, Masawa N, Fukasawa I. Natural selection in vertebrate evolution under genomic and biosphere biases based on amino acid content: Primitive vertebrate hagfish (*Eptatretsu burgeri*). Natural Science. 2013;**5**:221-227

[48] Saitou N, Nei M. The neighbor-joining method: A new method for reconstructing phylogenetic trees. Molecular Biology and Evolution. 1987;**4**:406-425

[49] Sorimachi K, Okayasu T, Ebara Y, Furuta E, Ohhira S. Phylogenetic position of *Xenoturubella Bocki* and Hemichordates *Balanoglossus carnosus* and *Saccoglossus kowalevskii* based on amino acid composition or nucleotide content of complete mitochondrial genomes. International Journal of Biology. 2014;**6**:82-94

[50] Sorimachi K, Okayasu T. Claasification of non-animals and invertevrates based on amino acid composition of complete mitochondrial genomes. International Journal of Biology. 2014:1-6

[51] Ward JH. Hierarchic grouping to optimize an objective function. Journal of the American Statistical Association. 1963;**58**:236-244

[52] Janvier P. Micro RNAs revive old views about jawless vertebrate divergence and evolution. Proceedings

of the National Academy of Sciences of the United States of America. 2010;**107**:19137-19138

[53] Sorimachi K. The most primitive extant ancestor of organisms and discovery of definitive evolutionary equations based on complete genome structures. Natural Science. 2018;**10**(9):338-369

[54] Lobry JR. Asymmetric substitution patterns in the two DNA strands of bacteria. Molecular Biology and Evolution. 1996;**13**:660-665

[55] Tillier ER, Collins RA. The contributions of replication orientation, gene direction, and signal sequences to base-composition asymmetries in bacterial genomes. Journal of Molecular Evolution. 2000;**50**:249-257

[56] Anderson S et al. Sequence and organization of the human mitochondrial genome. Nature. 1981;**290**:457-465

[57] Fonceca MM, Harris DJ, Posada D. The inversion of the control region in three mitogenomes pro vides further evidence for an asymmetric model of vertebrate mtDNA replication. PLoS One. 2014;**9**:e106654

[58] Seligmann H. Coding constraints modulate chemically spontaneous mutational replication gradients in mitochondrial genomes. Current Genomics;**13**:37-54

[59] Sorimachi K. Origine of life in the ocean: Direct derivation of mitochondria from primitive organisms based on complete genomes. Current Chemical Biology. 2015;**9**:23-35

[60] King N et al. The genome of the choanoflagellates Monosigarevicollis and the origin of metazoans. Nature. 2008;**451**:783-788

[61] Andersson SG et al. The genome sequence of *Rickettsia prowazekii* and the origin of mitochondria. Nature. 1998;**396**:133-140

[62] Sorimachi K. Codon evolution in double-stranded organelle DNA: Strong regulation of homonucleotides and their analog alternations. Natural Science. 2010;**2**:846-854

Chapter 3

Semantic Similarity in Cheminformatics

João D. Ferreira and Francisco M. Couto

Abstract

Similarity in chemistry has been applied to a variety of problems: to predict biochemical properties of molecules, to disambiguate chemical compound references in natural language, to understand the evolution of metabolic pathways, to predict drug-drug interactions, to predict therapeutic substitution of antibiotics, to estimate whether a compound is harmful, etc. While measures of similarity have been created that make use of the structural properties of the molecules, some ontologies (the Chemical Entities of Biological Interest (ChEBI) being one of the most relevant) capture chemistry knowledge in machine-readable formats and can be used to improve our notions of molecular similarity. Ontologies in the biomedical domain have been extensively used to compare entities of biological interest, a technique known as ontology-based semantic similarity. This has been applied to various biologically relevant entities, such as genes, proteins, diseases, and anatomical structures, as well as in the chemical domain. This chapter introduces the fundamental concepts of ontology-based semantic similarity, its application in cheminformatics, its relevance in previous studies, and future potential. It also discusses the existing challenges in this area, tracing a parallel with other domains, particularly genomics, where this technique has been used more often and for longer.

Keywords: semantic similarity, ontologies, ChEBI, prediction of molecule properties, databases

1. Introduction

With the unprecedented amount of data being generated today, we must start (and in some cases have already started) to rely on automatic systems to process, analyse, and understand all the scientific information that we produce. For some examples in chemistry, consider the number of drugs represented in DrugBank, which grew from 3909 in 2006 to 9688 [1], about 13% each year; the number of metabolites in the Human Metabolite Database grew from 2180 in 2007 to 114,100 in 2017 [2], approximately 39% per year (although at some point this database imported a large number of metabolites at once, artificially increasing this statistic); ChemSpider had 25 million compounds in 2010 [3] and now has 63 million (10% a year); and PubChem grew from 19 million compound structures in 2008 [4] to 96.5 million in August 2018 [5] (16% a year). These numbers usually grow exponentially [6], reflecting the fact that the amount of knowledge the scientific community produces is proportional to the amount of knowledge we discover.

With such high volumes of data, it is imperative that we categorise this information in ways that assist us in the tasks of consuming that information, specifically through categorisation schemas that abstract away the less useful details of reality and increase the manageability of this information. As we will see later in this chapter, ontologies can perform that goal: they are computational artefacts (files, tables in a database, etc.) whose goal is to encode real-world knowledge in machine-readable logical axioms that can be used by automatic systems to manipulate the knowledge inferred and potentially derivable from the data we have.

Furthermore, like most other scientific knowledge, chemistry ideas and notions are inferred from comparing entities and finding their similarities and differences. For instance, compound similarity has been used to (i) develop pharmacophores [7, 8], (ii) estimate whether a compound is harmful without in vivo experimentation [9], (iii) understand the evolution of metabolic pathways [10], (iv) predict adverse side effects of drugs [11], and (v) perform pharmacological profiling of compounds in drug design [12].

As we explore in this chapter, ontologies provide one way to measure similarity of chemistry entities (compounds, substances, mixtures, reactions, etc.), a technique known as ontology-based semantic similarity (shortened to semantic similarity in this chapter). This idea is already widely used in genomics and proteomics, but its full potential still needs to be brought over to other domains. While some research has successfully used this methodology in the cheminformatics domain (which we discuss below), there is still space for improvement and further methodological development.

In this chapter, we explore the ideas and concepts behind semantic similarity and chemistry ontologies, explore some past applications that use those concepts to further our knowledge of the chemical domain, and expose some limitations and challenges that this technique still needs to overcome for its whole potential to be released.

2. Measures of similarity in chemistry

Similarity, in its nature, is a notion that produces a number. In that sense, it is mathematical. However, chemical knowledge cannot be trivially reduced to mathematical form. For example, given two molecules, how should one compare them and assign a number to represent their similarity? And even if specific cases can be handled by humans, we still need an automatic way to perform comparison. However, to a certain extent, computers can only manipulate objects that can be represented mathematically (e.g., vectors) or as strings of characters (e.g., gene sequences, SMILES). But the algorithms that are used with these structures are context-free: they usually transform the structures without any knowledge of what they represent.

Many mechanisms exist to deal with this issue. For example, graph similarity can be used to find common substructures in two molecules as a basis for similarity calculations (see, e.g., [13, 14]), but these methods tend to be slow and computationally expensive. There is also the possibility to reduce a molecular structure into a *fingerprint*, which is a binary vector where each position represents the presence (with a 1) or absence (with a 0) of a certain feature in the structure. For example, the presence of a carboxyl group could be indicated with a 1 in some position of the vector. Similarity can then be computed by measuring the overlap in those vectors [15, 16].

These methods provide a high similarity value when the structures of the two molecules are high. Under the quantitative structure-activity relationship (QSAR)

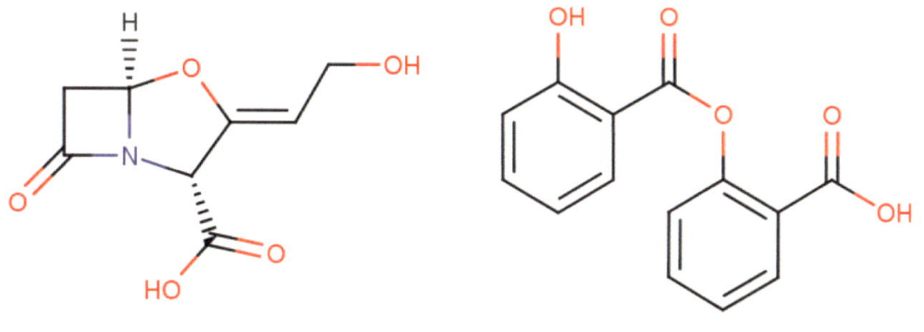

Figure 1.
Chemical structure of two semantically related compounds. The two molecular structures in the figure are quite different structures, and yet both present the same biological activity, namely, they inhibit β-lactamase enzymes.

premise, this means that, in general, two molecules with a high similarity score (as defined by these methods) tend to have similar biological role, similar chemical properties (such as melting point, optical parameters, and mass spectroscopy spectra), similar safety warnings, similar appearance, etc. But this is not always true. For instance, while L-amino acids are used to synthesise proteins, D-amino acids are much less frequent in nature, and their role is quite different [17]. From a biological point of view, they are distinct; however, to capture their structural differences, one needs to use three-dimensional methods, and even with that consideration, the structural similarity will be high, because both molecules have the same atoms and bonds. Another possibility includes simulation of docking with target proteins, but these methods are quite expensive computationally. Furthermore, not only can similar molecules perform different biological roles, different molecules can perform similar roles. For example, both clavulanic acid and salsalate are β-lactamase inhibitors, despite their different structures (see **Figure 1**).

Another way to measure similarity is by means of the semantics attached to the chemical compounds. Here, we use the term *semantics* to mean the knowledge that exists about a compound. This includes not only the structure of the molecule itself (e.g., the atomic connectivity, the number of oxygen atoms, the presence of triple bonds) but also other types of contextual knowledge, such as its chemical role (e.g., whether it is an electron donor, a solvent, or an explosive), biological role (e.g., whether it is a poison, a cofactor, or a vitamin), its applications (as a drug, fertiliser, fuel, etc.), its relationship to other molecules (such as being enantiomers, parent hydrides, etc.), and so on.

The difficulty with this is that knowledge is not directly machine-readable. Indeed, established facts have been traditionally published in plain text, which enables some humans to understand them; however, natural language processing techniques are not yet fully capable of converting scientific text into actionable formats (e.g., formats that allow automatic reasoning). Therefore, to enable the application of computerised processing power to knowledge manipulation, it is essential to find ways to represent knowledge in machine-readable formats.

3. Ontologies

Ontologies are the solution to this problem. An ontology is a representation of concepts from a domain of knowledge and the relationship between them and is usually visualised as a directed acyclic graph (DAG), where nodes are the concepts, edges are the relationships, and there are no cycles in the graph. See, for example,

Figure 2, a toy exampled based on a real-world ontology that encodes the fact that "acetate" is the conjugate base of "acetic acid" and that "acetic acid" is the conjugate acid of "acetate" and then organises these concepts in a hierarchy that contains concepts like "ion", "molecule", "organic acid", and "organic molecular entity", and ends up in the most generic "molecular entity" concept.

There are many ontologies whose purpose is to encode the chemical knowledge, but one of the most comprehensive and used is the ontology for Chemical Entities of Biological Interest (ChEBI) [18]. This ontology represents in a machine-readable format about 114 thousand concepts, including not only the chemical compounds but also their biological and chemical roles. Other ontologies that encode this or related domains include (*i*) Interlinking Ontology for Biological Concepts, (*ii*) Current Procedural Terminology, (*iii*) SNOMED CT, (*iv*) Chemical Information Ontology, and (*v*) Chemical Methods Ontology.

It is important to notice that, even though the notion of ontologies usually requires some logic concepts (such as axioms, predicates, etc.), some classification hierarchies are also sometimes named "ontologies". MeSH, the system used

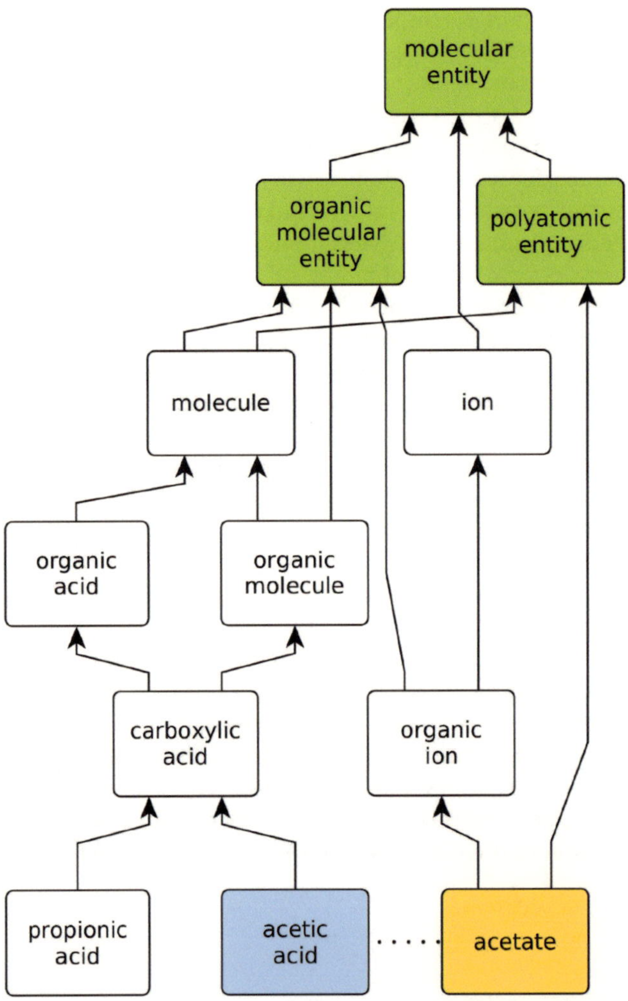

Figure 2.
A toy example of an ontology for chemical compounds, based on ChEBI. The ontology shows "is-a" relationships with solid lines, and a relationship between acid/base conjugates with a dotted line. The green shaded concepts are those that subsume both the yellow and the blue ones.

by PubMed to classify publications, is a hierarchy of concepts that possesses many of the same properties that ontologies do, namely, that it can be represented as a directed acyclic graph. However, one of the differences is that the relationship between two concepts does not always carry the same meaning. For example, "Head" is categorised under "Body Regions", and "Ear" is categorised under "Head", but while heads *are* body regions, ears *are not* heads; they are instead *parts* of the head. This illustrates the informality of MeSH: only one relationship type exists and it is used to express different notions. Another system in this category is the Anatomical Therapeutic Chemical (ATC) Classification System.

BioPortal [19], a repository of ontologies for the biomedical domain, contains a collection of 948 ontologies at the time of this writing. As an illustration of its magnitude, consider that 19 ontologies represent the concept "lidocaine". This reflects the effort being currently spent to represent human knowledge in machine-readable ontologies. In fact, while ontologies such as ChEBI are massive, BioPortal allows their users to submit new ontologies, even if small, focussed on a specific domain, and created with a specific application in mind other than pure knowledge representation (e.g., there is an ontology specific for cardiovascular drug adverse events, with 3 thousand concepts).

Other efforts have been set into place to aggregate ontologies in a single source of knowledge. For example, the Open Biological and Biomedical Ontology (OBO) Foundry [20] developed the OBO file format to represent ontologies and currently defines principles of quality for ontologies in biomedical domain that prescribe good practices for ontology development, such as being open, being reusable, being developed with collaboration in mind, containing both textual and logical definitions (for the benefit of both humans and machines), etc. They contain more than 200 ontologies as of this writing, 10 of which fully adhere to those principles (ChEBI being one of them). The OBO Foundry is tightly coupled with Ontobee [21], a web service that uses the principles of linked data to serve as a linked data server specifically targeted for ontologies and their concepts.

4. Semantic similarity

Using a formal representation of knowledge, computers are given the ability to manipulate concepts that are difficult to represent, in a way that preserves their "semantics". Ontologies provide the appropriate support for automatic manipulation of information. In this context, semantic similarity is a technique that assigns a numeric value to a pair of concepts based on the similarity of their meaning, extracted from the ontology.

For example, there is no directly obvious way to compare two roles. However, considering the illustration in **Figure 3**, it is possible to intuitively understand that, because both "hallucinogen" and "antifungal drug" are examples of "drugs", they are more similar than "hallucinogen" and "fossil fuel". This measure makes use of the meaning of the concepts, implicitly represented in the ontologies through the relations between the concepts. Ontologies function as a proxy for that meaning and enable its manipulation and ultimately comparison.

Several formulas and ideas have been proposed, implemented and tested in the past to compute semantic similarity. A full exposition on such measures and algorithms is beyond the scope of this chapter. The reader is encouraged to expand on this topic by reading works such as [22–25]. As such, the following is an abridged version of how ontology-based semantic similarity has been computed. In this discussion, consider the ontology in **Figure 3**.

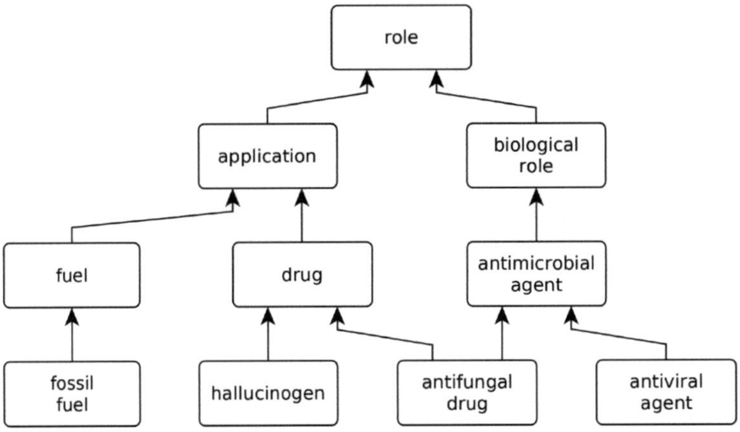

Figure 3.
A second toy example of an ontology representing chemical roles, also based on ChEBI.

Measures of similarity based on ontologies can roughly be divided into edge-based and node-based. An example of an edge-based measure is counting how many relations must be traversed to connect the two concepts being compared. Rada et al. [26] define distance as the number of edges in the smallest path between two nodes composed only of "is-a" relations. In this case, the distance between "hallucinogen" and "antimicrobial agent" would be three ("hallucino gen"→"drug"→"antifungal drug"→"antimicrobial agent"). While this type of approach is intuitive, it assumes that all nodes and all edges are equally important in terms of their semantics (e.g., that all edges weigh the same), which is generally not true in ontologies in life sciences. For instance, the "is-a" relation between "hallucinogen" and "drug" does not necessarily convey the same *amount of information* as the inverse "is-a" relation between "drug" and "antifungal drug".

One way to solve this is to introduce node-based methods, a technique that weighs nodes based on their *information content* (IC) [27]. The IC of a node is a numeric value based that reflects how informative its presence is and is calculated based on its frequency of use, since concepts that appear more frequently are generally less informative. The first formula proposed to measure IC was

$$\mathrm{IC}(c) = -\log f(c) \tag{1}$$

where $f(c)$ is the relative frequency with which the concept c and all its descendants appear in a corpus (in the example ontology, we can use the fraction of chemical concepts in ChEBI annotated as performing each of those roles). The intuition behind this idea is the following: consider a document (e.g., a scientific article) that uses the sentence "rodents have fur". The term "rodent" is used in such a way that every other concept that can be categorised under it also possesses the declared property. In fact, whenever a concept is used (in text, in logical axioms, etc.), it must be interpreted as including the set of all concepts recursively categorised under it.

The similarity between two concepts can be computed as the IC of the *most informative common ancestor* (usually abbreviated as MICA) between them

$$\mathrm{sim}_{\mathrm{Resnik}}(c_1, c_2) = \mathrm{IC}(\mathrm{MICA}(c_1, c_2)). \tag{2}$$

This idea has been iterated upon with some additions and adaptations.

- The IC measure can be normalised so that it ranges from 0.0 to 1.0 (originally, the measure is unbounded above);

- The IC measure has been computed from multiple sources, such as (*i*) text corpora (as in the original), (*ii*) frequency of usage of the ontology concepts in external sources [28], or (*iii*) the ontology itself, where frequency can be computed based on the number of descendants (direct or indirect) of a concept [29], the number of leaf descendants of a concept [30], or other topological properties of the graph representation of the ontology [31].

- The semantic similarity measure itself can be normalised. Notice that the original measure gives the same similarity to the pair "application"/"biological role" (both generic concepts) and "fossil fuel"/"antiviral agent", which goes against the intuition that the first pair should be more similar. Lin [32] uses this idea to define

$$\mathrm{sim}_{\mathrm{Lin}}(c_1, c_2) = \frac{2 \cdot \mathrm{IC}(\mathrm{MICA}(c_1, c_2))}{\mathrm{IC}(c_1) + \mathrm{IC}(c_2)} \quad ; \qquad (3)$$

- The notion of shared information content (originally measured as the information content of the MICA of the two concepts) has also been tuned to take into account the fact that concepts can have multiple parents [33], which is necessary in many life science fields since it is in the nature of biomedical ontologies that some concepts are categorised under multiple parents, (see https://github.com/lasige-BioTM/DiShIn for an example of software that computes this type of measure) or the fact that ontologies have disjointness axioms that encode the fact that two concepts cannot share any descendants [34], also important because life science ontologies, and especially chemistry ones, make use of those types of axioms [35].

- The way to measure shared information content has also been completely re-implemented to use not the IC of the most informative common ancestor but a metric based on the set of all ancestors of the concepts [36].

These measures are able to compare one concept with another. It is also possible to compare sets of concepts. For this, one takes the matrix of pairwise similarities between concepts in the first set and concepts in the second set and mathematically manipulates it to produce a single number, taking, for example, the average, the maximum, or the "best match average", an approach that averages the highest values in each row and column [22]. There are other approaches that convert a set of concepts into the set of all their ancestors and take the intersection of those sets as a measure of similarity (two examples are simUI and simGIC [22]).

Finally, there is a difference in measuring the *similarity* or the *relatedness* between concepts. Similarity is a term that is generally applied to the notion that two concepts are "alike" and is usually computed based on "is-a" hierarchies; relatedness is more general: two related concepts can be related based on their categorisation on a hierarchy or on any number of other non-hierarchical relations. This distinction is important in chemistry, and ChEBI in particular, since many chemistry concepts are related via relations such as "has-role", "has-part", "is-enantiomer-of", etc.

Notice that when nothing is known about a chemical compound other than its structure, semantic methods can still be used, because one of the ways ontologies (especially ChEBI) classify molecules is based on their structure. For example, ChEBI has a concept "carboxylic acid" which is an ancestor of all molecules that have one or more carboxylic acid groups (e.g., benzoic acid, all amino acids, all penicillins, etc.). This, however, is not conceptually different from measuring structural similarity, and such a setting would lack the enrichment provided by other types of knowledge (e.g., the knowledge of the chemical and biological roles of the molecule).

5. Applications

Since 2003, when Lord et al. [28] introduced the idea of ontology-based semantic similarity in the gene ontology (GO), several results have been achieved using this technique, proving beyond doubt that it is sound and useful and has real-life applications. In genomics and proteomics, semantic similarity based on GO has been used to (i) cluster proteins [37], (ii) find protein-protein interactions [38], (iii) interpret microarray data [39], (iv) predict protein functions [40], (v) prioritise candidate disease genes [41], etc. Other uses outside GO include predicting disease-related phenotypes [42] and predicting clinical diagnosis from a set of phenotype abnormalities [43].

The uses in chemistry-related areas have been scarce, but nonetheless existing and with real-world applications. We collected three research studies of semantic similarity in cheminformatics, which show its use in this area.

5.1 Predict biochemical properties of molecules

In 2010, ontology-based semantic similarity was applied to ChEBI [44] using a methodology named Chym. Chym shows for the first time that semantic similarity is useful in biomedical chemistry, by applying these ideas to predict whether a molecule (i) is capable of crossing the blood brain barrier, (ii) is a substrate of the P-glycoprotein, and (iii) binds to an oestrogen receptor. These properties are at least partially intrinsically related to the three-dimensional structure of the molecules and also of the proteins that perform the biochemical role in the organism. However, the work shows that structural similarity alone can be improved if it is coupled with semantic similarity.

Chym used daylight fingerprints for structural similarity and simUI and simGIC for semantic similarity, using ChEBI as the ontology. For all the three properties mentioned above, Chym was able to clearly outperform what were then the state-of-the-art prediction techniques for those properties.

Notice that this means that the two ideas presented here, structural similarity and semantic similarity, are not orthogonal and can be applied simultaneously with good results. This is not surprising, as ontologies can complement the knowledge that can be inferred form the structure alone, without needing to resort to wet-lab experiments.

5.2 Disambiguate chemical compound references in natural language

As the amount of textual chemistry information increases, particularly in the form of drug leaflets, articles, patents, and other types of communications, the need to develop mechanisms to automatically read these texts and extract tractable information from them increases as well. In this context, named entity recognition is a text mining task whose goal is to identify the entities mentioned in text.

There have been many attempts to create such systems in the chemical domain (see, e.g., the review [45]). In one of those attempts [46], semantic similarity has been used to improve the precision of existing methodologies by successfully identifying some false positives and removing them from the final result set. The idea of that work is that, within a scope of text (e.g., a sentence or a paragraph), chemical entities mentioned in that scope share some degree of semantic similarity that is higher than average. When entity recognition algorithms offer more than one possible ChEBI identifier for an excerpt of text, similarity with other ChEBI concepts can be used to disambiguate which is the correct entity.

5.3 Drug repurposing

Drug repurposing is the process by which drug that have therapeutic application are computationally tested to find other therapeutic applications. This reduces costs and improves the drug development pipeline and as such is important for the pharmaceutical industry.

The work presented in [47] couples similarity between the three-dimensional molecular structure with semantic similarity between the drug targets to find new indications for known drugs. The ontology used here is not a chemistry-specific one, but GO.

The main methodology of this work was:

1. Select a drug d and a potential target protein p.

2. Find drugs similar to this one (up to a threshold) with a structural similarity measure. Store these structural similarity values in a vector X str = $(d\,1, d\,2, \ldots, d\,m)$.

3. For each similar drug d_i, find its interacting proteins, compare them with p using GO-based semantic similarity, and sum the results. Call this value p_i. We have now a vector X sem = $(p\,1, p\,2, \ldots, p\,m)$.

4. The drug-protein association is assigned a score that depends on the correlation between the vectors X_{str} and X_{sem}. For a set of N proteins, each drug was then assigned a vector of N drug-protein association values, called the drug's "expression profile".

5. The drug-drug similarity measure was computed based on the correlation between the "expression profiles" of the two drugs.

The similarity between drugs was then used to construct a network of similarities, where clusters of highly connected drugs were indicative of potential drug repurposing.

A related work [48] also uses semantic similarity to predict drug-protein interaction. In this work, probabilistic similarity logic is used to construct models that are based on a notion of "similarity triads": triples of the form "drug-target-drug" with similar drugs or "target-drug-target" with similar targets. The whole work was based on the assumption that similar targets tend to interact with the same drug and similar drugs tend to interact with the same target. Here, several protein similarity methods (including semantic similarity based on GO) and drug similarity method (including semantic similarity based on ATC) were used to build a probabilistic model that predicts whether drugs and proteins interact.

6. Challenges and future work

Semantic similarity in cheminformatics has been slow to keep with the pace of equivalent research in other life science fields, such as genomics and proteomics. We posit that this is in some ways related to general and specific challenges associated with the application of this methodology in chemistry.

First, the state of ontology development and the more general knowledge representation area is very active, specifically in the biomedical fields. This means that many people have the motivation to develop their own ontology, with specific views of the reality embedded in it. However, as many people create their own knowledge representation artefacts, many different ontologies start to appear that overlap in domain, which means that it is not always obvious which ontology (or ontologies) to choose for a specific goal. Furthermore, these ontologies are not easy to reconcile, because they encode different and disjoint points of view. While efforts have been made to attenuate this problem, such as ontology matching (the process by which ontologies of the same domain are automatically merged into a single ontology) and the establishment of community standards (in chemistry, e.g., it is standard practice to reuse ChEBI concepts rather than create new concepts in new ontologies), the problem still persists.

Second, metrics of semantic similarity have been mostly developed and tested in the fields of natural language processing and genomics/proteomics. While these seem to have good enough results when used with ChEBI, we still do not know if they are the most adequate measures in this domain. Ferreira et al. [34] developed and validated a measure on the chemical domain, but more work needs to be done in this area. In particular, what role should the non-hierarchical relationship types ("is-enantiomer-of", "is-conjugate-acid-of", etc.) have in semantic similarity?

The third challenge is one of similarity profiles. It is not always obvious which details or properties of a molecule should be used for comparing. Should a pair of chemical compounds that differ only in the presence of an oxygen atom (e.g., methane vs. methanol) be more similar than a pair of molecules that differ only in charge (e.g., NO_2 vs. NO_2^-) or only in their three-dimensional conformation (e.g., L-serine vs. D-serine)? This problem must be solved based on context: determining what the similarity measure will be used for and then deciding which features are important. This includes deciding, for example, which relationship types should be taken into account, how to weight them, etc. Maggiora et al. [49] touch on the fact that chemoinformaticians and medicinal chemists typically perceive similarity differently and we need to find ways to capture those differences in actionable measures of similarity.

The fourth challenge is the necessity of taking into account multiple domains of knowledge: drugs interact with proteins, treat and cause diseases, are produced by different methods (industrial or otherwise), have side effects, participate in metabolic reactions, etc. These concepts from other domains can also be compared semantically (many are even already represented in appropriate ontologies, including diseases, proteins, types of molecular interaction, manufacturing procedures, side effects, and pathways). The question now is how to take advantage of these other ontologies in order to implement a useful and accurate measure of chemical similarity. This issue is even related to the previous one, since by tuning the weight of these other domains, we can create new profiles of similarity more pertinent to some goals than others.

Another challenge is the absence of a standardised way to *validate* the measures that are proposed. In practice, for each new measure being proposed by some research group, that same group validates the new measure by comparing them with previous ones or by using it to show that the new measure can find hidden knowledge in some dataset. However, the *ad hoc* way these validations are performed means that frequently the measures are neither comparable nor interchangeable and that they can only be used for the goal used to validate them. Thus, a general but useful validation strategy should also be developed to bring cohesion to this field.

7. Conclusion

This chapter introduces the ideas behind ontology-based semantic similarity measures, how they are applied in life sciences, and some of their uses in chemistry-related research endeavours. The main idea that we exposed is that these methods, having been used in other biomedical fields, can help cheminformatics in several fronts. We described three applications of where this methodology has been applied directly in cheminformatics research efforts and expect that this number grows as more people are exposed to this idea and its use cases.

We also exposed some of the future challenges in this area, which can serve as a starting point to anyone wishing to improve on the work already published, and provided general guidelines that should be taken into account for the further improvement of cheminformatics as a scientific field. In particular, we emphasise the need to explore the multidomain potential in semantic similarity, as well as the need to standardise the ways to validate measures of semantic similarity.

Acknowledgements

This work was supported by FCT through funding of DeST: Deep Semantic Tagger project, ref. PTDC/CCI-BIO/28685/2017 (http://dest.rd.ciencias.ulisboa.pt./) and LASIGE Research Unit, ref. UID/CEC/00408/2019.

Abbreviations

ATC	anatomical therapeutic chemical classification system
ChEBI	chemical entities of biological interest
DAG	directed acyclic graph
GO	gene ontology
IC	information content
MeSH	medical subject headings
MICA	most informative common ancestor
OBO	Open Biological and Biomedical Ontology
QSAR	quantitative structure-activity relationship
simGIC	similarity of graphs with information content
simUI	similarity with union and intersection
SMILES	simplified molecular-input line-entry system
SNOMED CT	systematised nomenclature of medicine—clinical terms

Author details

João D. Ferreira* and Francisco M. Couto
LASIGE, Departamento de Informática, Faculdade de Ciências, Universidade de Lisboa, Portugal

*Address all correspondence to: jdferreira@fc.ul.pt

References

[1] Wishart DS, Feunang YD, Guo AC, Lo EJ, Marcu A, Grant JR, et al. DrugBank 5.0: A major update to the DrugBank database for 2018. Nucleic Acids Research. 2017;**46**(D1):D1074-D1082. Available from: 10.1093/nar/gkx1037

[2] Wishart DS, Feunang YD, Marcu A, Guo AC, Liang K, Vázquez-Fresno R, et al. HMDB 4.0: The human metabolome database for 2018. Nucleic Acids Research. 2017;**46**(D1):D608-D617. Available from: 10.1093/nar/gkx1089

[3] Pence HE, Williams A. ChemSpider: An online chemical information resource. Journal of Chemical Education. 2010;**87**(11):1123-1124. Available from: 10.1021/ed100697w

[4] Bolton EE, Wang Y, Thiessen PA, Bryant SH. Chapter 12—PubChem: Integrated platform of small molecules and biological activities. In: Wheeler RA, Spellmeyer DC, editors. Annual Reports in Computational Chemistry. Vol. 4. Amsterdam, The Netherlands: Elsevier; 2008. pp. 217-241. Available from: http://www.sciencedirect.com/science/article/pii/S1574140008000121

[5] Kim S, Chen J, Cheng T, Gindulyte A, He J, He S, et al. PubChem 2019 update: Improved access to chemical data. Nucleic Acids Research. 2018;**47**(D1):D1102-D1109. Available from: 10.1093/nar/gky1033

[6] Larsen PO, von Ins M. The rate of growth in scientific publication and the decline in coverage provided by science citation index. Scientometrics. 2010;**84**(3):575-603. Available from: 10.1007/s11192-010-0202-z

[7] Penzotti JE, Lamb ML, Evensen E, Grootenhuis PDJ. A computational ensemble pharmacophore model for identifying substrates of P-glycoprotein. Journal of Medicinal Chemistry. 2002;**45**(9):1737-1740

[8] Fukunishi Y, Mikami Y, Takedomi K, Yamanouchi M, Shima H, Nakamura H, et al. Classification of chemical compounds by protein-compound docking for use in designing a focused library. Journal of Medicinal Chemistry. 2006;**49**(2):523-533

[9] Richard AM, Gold LS, Nicklaus MC. Chemical structure indexing of toxicity data on the internet: Moving toward a flat world. Current Opinion in Drug Discovery & Development. 2006;**9**(3):314-325

[10] Tohsato Y, Nishimura Y. Metabolic pathway alignment based on similarity between chemical structures. Information and Media Technologies. 2008;**3**(1):191-200

[11] Huang LC, Wu X, Chen JY. Predicting adverse side effects of drugs. BMC Genomics. 2011;**12**(5):S11. Available from: 10.1186/1471-2164-12-S5-S11

[12] Nikolic K, Mavridis L, Djikic T, Vucicevic J, Agbaba D, Yelekci K, et al. Drug design for cns diseases: polypharmacological profiling of compounds using cheminformatic, 3D-QSAR and virtual screening methodologies. Frontiers in Neuroscience. 2016;**10**:265. Available from: https://www.frontiersin.org/article/10.3389/fnins.2016.00265

[13] Raymond JW, Gardiner EJ, Willett P. Heuristics for similarity searching of chemical graphs using a maximum common edge subgraph algorithm. Journal of Chemical Information and Computer Sciences. 2002;**42**(2):305-316. PMID: 11911700. Available from: 10.1021/ci010381f

[14] Gillet VJ, Willett P, Bradshaw J. Similarity searching using reduced graphs. Journal of Chemical Information and Computer Sciences. 2003;**43**(2):338-345. PMID: 12653495. Available from: 10.1021/ci025592e

[15] Rogers D, Hahn M. Extended-connectivity fingerprints. Journal of Chemical Information and Modeling. 2010;**50**(5):742-754. PMID: 20426451. Available from: 10.1021/ci100050t

[16] Daylight Chemical Information Systems, Inc. Daylight Theory Manual. Daylight Headquarters; 2011 [Online]. Available from: https://www.daylight.com/dayhtml/doc/theory/ [Accessed: 19 June 2019]

[17] Wolosker H, Dumin E, Balan L, Foltyn VN. D-amino acids in the brain: D-serine in neurotransmission and neurodegeneration. The FEBS Journal. 2008;**275**(14):3514-3526

[18] Hastings J, Owen G, Dekker A, Ennis M, Kale N, Muthukrishnan V, et al. ChEBI in 2016: Improved services and an expanding collection of metabolites. Nucleic Acids Research. 2015;**44**(D1):D1214-D1219. Available from: 10.1093/nar/gkv1031

[19] Whetzel PL, Noy NF, Shah NH, Alexander PR, Nyulas C, Tudorache T, et al. BioPortal: Enhanced functionality via new web services from the National Center for biomedical ontology to access and use ontologies in software applications. Nucleic Acids Research. 2011;**39**(suppl 2):W541-W545. Available from: 10.1093/nar/gkr469

[20] Smith B, Ashburner M, Rosse C, Bard J, Bug W, Ceusters W, et al. The OBO foundry: Coordinated evolution of ontologies to support biomedical data integration. Nature Biotechnology. 2007;**25**(11):1251

[21] Xiang Z, Mungall C, Ruttenberg A, He Y. Ontobee: A linked data server and browser for ontology terms. In: Proceedings of the 2nd International Conference on Biomedical Ontology; 2011

[22] Pesquita C, Faria D, Falcao AO, Lord P, Couto FM. Semantic similarity in biomedical ontologies. PLoS Computational Biology. 2009;**5**(7):e1000443

[23] Harispe S, Sánchez D, Ranwez S, Janaqi S, Montmain J. A framework for unifying ontology-based semantic similarity measures: A study in the biomedical domain. Journal of Biomedical Informatics. 2014;**48**:38-53

[24] Sánchez D, Batet M. Semantic similarity estimation in the biomedical domain: An ontology-based information-theoretic perspective. Journal of Biomedical Informatics. 2011;**44**(5):749-759. Available from: http: //www.sciencedirect.com/science/article/pii/S1532046411000645

[25] Ferreira JD. Semantic similarity across biomedical ontologies [PhD thesis]. Universidade de Lisboa; 2016. Available from: http://hdl.handle.net/10451/25070

[26] Rada R, Mili H, Bicknell E, Blettner M. Development and application of a metric on semantic nets. IEEE Transactions on Systems, Man, and Cybernetics. 1989;**19**(1):17-30

[27] Resnik P. Semantic similarity in a taxonomy: An information-based measure and its application to problems of ambiguity in natural language. Journal of Artificial Intelligence Research. 1999;**11**:95-130

[28] Lord PW, Stevens RD, Brass A, Goble CA. Investigating semantic similarity measures across the gene ontology: The relationship between sequence and annotation. Bioinformatics. 2003;**19**(10):1275-1283.

Available from: 10.1093/bioinformatics/btg153

[29] Seco N, Veale T, Hayes J. An intrinsic information content metric for semantic similarity in WordNet. In: Proceedings of the 16th European Conference on Artificial Intelligence; ECAI'04; Amsterdam, The Netherlands, The Netherlands: IOS Press; 2004. pp. 1089-1090. Available from: http://dl.acm.org/citation.cfm?id=3000001.3000272

[30] Sánchez D, Batet M, Isern D. Ontology-based information content computation. Knowledge-Based Systems. 2011;**24**(2):297-303. Available from: http://www.sciencedirect.com/science/article/pii/S0950705110001619

[31] Zhou Z, Wang Y, Gu J. A new model of information content for semantic similarity in WordNet. In: 2008 Second International Conference on Future Generation Communication and Networking Symposia; vol. 3; 2008. pp. 85-89

[32] Lin D. An information-theoretic definition of similarity. In: Proceedings of the Fifteenth International Conference on Machine Learning; ICML '98; San Francisco, CA, USA: Morgan Kaufmann Publishers Inc.; 1998. pp. 296-304. Available from: http://dl.acm.org/citation.cfm?id=645527.657297

[33] Couto FM, Silva MJ. Disjunctive shared information between ontology concepts: Application to gene ontology. Journal of Biomedical Semantics. 2011;**2**(1):5. Available from: 10.1186/2041-1480-2-5

[34] Ferreira JD, Hastings J, Couto FM. Exploiting disjointness axioms to improve semantic similarity measures. Bioinformatics. 2013;**29**(21):2781-2787. Available from: 10.1093/bioinformatics/btt491

[35] Hastings J, de Matos P, Dekker A, Ennis M, Muthukrishnan V, Turner S, et al. Modular extensions to the ChEBI ontology. In: Cornet R, Stevens R, editors. Proceedings of the 3rd International Conference on Biomedical Ontology (ICBO 2012); KR-MED Series, Graz, Austria; 21-25 July 2012; vol. 897 of CEUR Workshop Proceedings; CEUR-WS.org; 2012. Available from: http: //ceur-ws.org/Vol-897/poster 7.pdf

[36] Batet M, Sánchez D, Valls A. An ontology-based measure to compute semantic similarity in biomedicine. Journal of Biomedical Informatics. 2011;**44**(1):118-125. Ontologies for Clinical and Translational Research. Available from: http: //www.sciencedirect.com/science/article/pii/S1532046410001346

[37] Brameier M, Wiuf C. Co-clustering and visualization of gene expression data and gene ontology terms for Saccharomyces cerevisiae using self-organizing maps. Journal of Biomedical Informatics. 2007;**40**(2):160-173. Available from: http: //www.sciencedirect.com/science/article/pii/S153204640600061X

[38] Zhang SB, Tang QR. Protein-protein interaction inference based on semantic similarity of gene ontology terms. Journal of Theoretical Biology. 2016;**401**:30-37

[39] Yang D, Li Y, Xiao H, Liu Q, Zhang M, Zhu J, et al. Gaining confidence in biological interpretation of the microarray data: The functional consistence of the significant GO categories. Bioinformatics. 2007;**24**(2):265-271. Available from: 10.1093/bioinformatics/btm558

[40] Jiang Y, Oron TR, Clark WT, Bankapur AR, D'Andrea D, Lepore R, et al. An expanded evaluation of protein function prediction methods shows an improvement in accuracy. Genome

Biology. 2016;**17**(1):184. Available from: 10.1186/s13059-016-1037-6

[41] Liu B, Jin M, Zeng P. Prioritization of candidate disease genes by combining topological similarity and semantic similarity. Journal of Biomedical Informatics. 2015;**57**:1-5. Available from: http: //www.sciencedirect.com/science/article/pii/S1532046415001458

[42] Xue H, Peng J, Shang X. Predicting disease-related phenotypes using an integrated phenotype similarity measurement based on HPO. BMC Systems Biology. 2019;**13**(2):34. Available from: 10.1186/s12918-019-0697-8

[43] Köhler S, Schulz MH, Krawitz P, Bauer S, Dlken S, Ott CE, et al. Clinical diagnostics in human genetics with semantic similarity searches in ontologies. The American Journal of Human Genetics. 2009;**85**(4):457-464. Available from: http: //www.sciencedirect.com/science/article/pii/S0002929709003991

[44] Ferreira JD, Couto FM. Semantic similarity for automatic classification of chemical compounds. PLoS Computational Biology. 2010;**6**(9):1-11. Available from: 10.1371/journal.pcbi.1000937

[45] Eltyeb S, Salim N. Chemical named entities recognition: A review on approaches and applications. Journal of Cheminformatics. 2014;**6**(1):17. Available from: 10.1186/1758-2946-6-17

[46] Lamurias A, Ferreira JD, Couto FM. Improving chemical entity recognition through h-index based semantic similarity. Journal of Cheminformatics. 2015;**7**(1):S13. Available from: 10.1186/1758-2946-7-S1-S13

[47] Tan F, Yang R, Xu X, Chen X, Wang Y, Ma H, et al. Drug repositioning by applying "expression profiles" generated by integrating chemical structure similarity and gene semantic similarity. Molecular BioSystems. 2014;**10**:1126-1138. Available from: 10.1039/C3MB70554D

[48] Fakhraei S, Raschid L, Getoor L. Drug-target interaction prediction for drug repurposing with probabilistic similarity logic. In: Proceedings of the 12th International Workshop on Data Mining in Bioinformatics; BioKDD '13. New York, NY, USA: ACM; 2013. pp. 10-17. DOI: 10.1145/2500863.2500870

[49] Maggiora G, Vogt M, Stumpfe D, Bajorath J. Molecular similarity in medicinal chemistry. Journal of Medicinal Chemistry. 2014;**57**(8):3186-3204. PMID: 24151987. Available from: 10.1021/jm401411z

Chapter 4

Molecular Electrostatic Potential and Chemometric Techniques as Tools to Design Bioactive Compounds

Marcos Antônio B. dos Santos, Luã Felipe S. de Oliveira, Antônio Florêncio de Figueiredo, Fábio dos Santos Gil, Márcio de Souza Farias, Heriberto Rodrigues Bitencourt, José Ribamar B. Lobato, Raimundo Dirceu de P. Farreira, Sady Salomão da S. Alves, Edilson Luiz C. de Aquino and José Ciríaco-Pinheiro

Abstract

In this chapter, firstly, we briefly review aspects of the approximation of quantum chemistry, molecular electrostatic potential (MEP), and chemometrics techniques, which are accredited as important tools in the development of chemical science and are frequently used in the study and design of bioactive compounds. Ultimately, we use MEP and pattern recognition (PR) techniques as tools to design nitrofuran compounds with biological activity against *Trypanosoma cruzi* (*T. cruzi*). PR models (PCA, HCA, KNN, SDA, and SIMCA) were constructed and demonstrated that 23 nitrofurans can be classified into two classes or groups: more active and less active according to their degrees of activity against *T. cruzi*. Properties such as charge on the N atom of the nitro group (QN1); the difference between the highest occupied molecular orbital (HOMO) energy and the lowest unoccupied molecular orbital (LUMO) energy (GAP energy); molecular representation of structure based on electron diffraction code of signal 5, unweighted (Mor05u); and Moriguchi water–octanol partition coefficient (MlogP) are responsible for the classification into more active and less active studied nitrofurans. It is interesting to notice that these properties represent three distinct classes of interactions between the nitrofurans and the biological receptor: electronic (QN1 and GAP energy), steric (Mor05u), and hydrophobic (MlogP). The results of the application of PR models on the validation set evidenced two nitrofuran compounds (compounds **25** and **30**) as more promising for synthesis and biological assays, which in the future can be used to validate our PR models.

Keywords: molecular electrostatic potential, chemometric techniques, pattern recognition techniques, chemoinformatics, design of bioactive compounds

1. Introduction

Reports of theoretical bases of MEP and the development of efficient computational methods state that MEP has become an important reactivity index in studies of a large variety of molecular interactions [1]. The usefulness of this theoretical approach in studies and interpretation of chemical, biochemical, and related phenomena is well documented [2–18].

Chemometrics is a discipline that collects mathematical, statistical, information theory, and computer science tools to deal with complex chemical data [19–22]. PR techniques were introduced in the chemistry, at the beginning of the 1970s, to analyze various types of spectroscopic data. Since then, PR became part of chemometrics and has been an excellent tool to aid in the interpretation of chemical data to obtain relevant information in different application sectors of chemical science [19, 20]. PR techniques are especially useful for the classification of objects into discrete classes on the basis of measured features. A set of characteristic features of an object is considered as an abstract pattern that contains information about a not directly measured property of the object [19].

The MEP and PR techniques have been used as independent strategies in the study of active compounds and lead to the proposal of new molecules for synthesis and biological testing. The joint applications of these powerful tools were described carefully, to unravel the structure-activity relationship of bioactive compounds, consequently proposing new molecules. Therefore, a more intense exploration of its potentials is needed in order to design biologically active compounds.

The design of molecules with a desired property is one of the objectives of chemoinformatics. In this chapter, we present a study of the application of MEP and PR techniques to design nitrofuran compounds with potential activity against *T. cruzi*. In the first step of our study, MEP maps will be used in an attempt to identify the key structural features of nitrofuran compounds that are necessary for their activities and investigate their probable interactions with a molecular receptor through recognition in a biological process. Subsequently, PR techniques are used to construct models that will be applied later to a forecast set constructed with the accumulated perceptions in the MEP studies.

2. MEP and chemometrics techniques as tools for the design of bioactive compounds: a brief review

According to the literature, MEP [1, 3] has been a tool of quantum chemistry used by researchers for several decades to study and understand the relationships between structure and activity of molecules. Among the papers that point out the importance of this tool in the matter, and consequently in the planning of bioactive compounds, we can mention those reported by Bernardinelli et al. [23] and by Jefford et al. [24].

Another tool, in the form of a set of techniques has been used emphatically over the years in the understanding of the structure-activity relationship of molecules is Chemometrics [25–27]. This set of techniques has also enables the planning of new biologically active compounds, and most of the developed research is focused on the construction of QSAR (quantitative structure-activity relationship) models.

The combination of MEP and chemometrics as tools for designing new bioactive compounds has almost always been focused on the elaboration of quantitative models, for example, the CoMFA methodology [28]. This methodology was developed in the late 1980s by Cramer et al. [29]. Its application is richly extensive and recently it has been used in several studies of structure–activity relationships of bioactive compounds. Chatbar et al. conducted a study of triazine morpholino derivatives as mTOR inhibitors for the treatment of breast cancer [30]. Pourbasheer et al. performed 3D-QSAR and 2D-QSAR analyses on the series of compounds hepatitis C virus NS5B polymerase inhibitors [31]. Cramer applied the CoMFA methodology for a large majority of 116 biological targets and obtained acceptable 3D-QSAR models [32]. Cramer et al. introduced in the literature a novel alignment methodology for training or test set structures in 3D-QSAR [33]. Dong et al. performed QSAR analyses of aromatic heterocycle thiosemicarbazone analogues for finding novel tyrosinase inhibitors [34]. Dong et al. built 3D-QSAR models of dabigatran analogues as thrombin inhibitors [35]. Ding et al. performed 3D-QSAR models of 6-aryl-5-cyano-pyrimidine derivatives to explore the structure requirements of LSD1 inhibitors [36].

Applications of MEP to investigate the key features of compounds that are necessary for their biological activities and thus proposing new derivatives as well as the construction of chemometric models as indicative of the most promising among the new derivatives for syntheses and biological assays were reported by us in literature [37–43]. Pinheiro et al. stated the use of MEP and partial least squares regression (PLS) method in the design of new artemisinin derivatives with activities against *Plasmodium falciparum* [37]. Cardoso et al., using MEP maps and multivariate QSAR, designed new artemisinin derivatives with antimalarial activity [38]. Ferreira et al., through MEP maps and multivariate analysis, designed antimalarial artemisinins [39]. Figueiredo et al. designed new derivatives of dispiro-1,2,4-trioxolones with activity against falciparum malaria [40]. Carvalho et al., through maps of MEP and pattern recognition methods, proposed new artemisinin derivatives with activity against *Leishmania donovani* [41]. Barbosa et al. used MEP maps and pattern recognition techniques to plan new derivatives of artemisinin anticancer HepG2 [42]. Cristino et al. proposed new derivatives of 10-substituted Deoartemisinis with activity against *P. falciparum* [43] through the use of MEP maps and pattern recognition techniques.

3. MEP and PR techniques as tools to design nitrofuran compounds with biological activity against *T. cruzi*

3.1 Computational

3.1.1 Biological recognition process ligand/receptor through the molecular electrostatic potential

The MEP is also suitable for analyzing processes based on the “recognition” of one molecule by another as in drug-receptor and enzyme-substrate interactions, because it is through their potentials that the two species first “see” each other [2, 3, 44–46].

MEP for the electronic density is a very useful property for understanding the site of electrophilic attack and nucleophilic reactions as well as the hydrogen bonding interactions [46]. The MEP at a given point (x, y, z) in the vicinity of a molecule is defined in terms of the interaction energy between the electrical charge generated from the molecule’s electrons and nuclei and a positive charge test (a proton) located at $\vec{r}$. Being a real physical property, MEP can be determined experimentally by diffraction or by computational tools [3]. For the studied nitrofuran molecules, the MEP values were computed through Eq. (1) [45]

$$V(\vec{r}) = \sum_{j=1}^{K} \frac{Z_j}{\left|\vec{R}_j - \vec{r}\right|} - \int \frac{\rho(\vec{r}\,')d\vec{r}\,'}{\left|\vec{r}\,' - \vec{r}\right|} \quad (1)$$

where K is the number of nuclei with charges Z_j, located at position R_j and $\rho(\vec{r})$ is the electronic charge density. The first term on the right side of Eq. (1) represents the contribution of the nuclei, which is positive; the second term brings in the effect of the electrons, which is negative. In the investigation of the reactive sites of nitrofuran compounds, the MEP was evaluated through of the HF/6-31G method.

3.1.2 RP techniques

In this section, we will make a brief presentation of the PR techniques used in this chapter. A deeper and detailed description of these matters can be found elsewhere [47–66].

3.1.2.1 Principal component analysis (PCA) technique

When computing large multivariate data, it is mandatory to find and reduce unknown data trends using exploratory tools. The main idea of the PCA technique is to reduce the dimensionality of a data set consisting of large numbers of interrelated variables while retaining the variation present in the data set as much as possible. This can be achieved by transforming them into a new set of variables, the PCs, which are uncorrelated and ordered so that the first few retain most of the variation present in all of the original variables. As the final result, the PCA technique performs the selection of a small number of variables (molecular properties) considered better related to the dependent property or feature [67], in this study, the biological activity against *T. cruzi*.

3.1.2.2 Hierarchical cluster analysis (HCA) technique

This technique has become, together with PCA, another important tool in pattern recognition [67]. The purpose of using it is to display the data in such a way as to emphasize its natural clusters and patterns in a two-dimensional space. The results are presented as dendrograms. In HCA technique, the distances between objects or variables are calculated and computed through the similarity index which ranges from zero, that is, no similarity and large distance among objects, to one, for identical objects.

3.1.2.3 K-nearest neighbor (KNN) technique

The KNN technique [67] classifies the objects based on distance comparison among them. The multivariate Euclidean distances between every pair of objects with known class membership are calculated. The closest K objects are used to build the model. The optimal K is determined by cross-validation applied to the training set objects. The classification of a test object is determined based on the multivariate distance of this object with respect to the K objects in the training set. In this technique no assumption is made about the size and shape of the training set classes.

3.1.2.4 Stepwise discriminant analysis (SDA) technique

This technique separates objects from distinct populations and allocates new objects into populations previously defined. It uses a stepwise procedure in which, at each step, the most powerful variable is entered into the discriminant function. The SDA technique is anchored in the F-test for the significance of variables and at each step selects a variable based on its significance, and, after several steps, the most significant variables are extracted from the set in question [20, 68].

3.1.2.5 Soft independent modeling of class analogy (SIMCA) technique

This SIMCA technique develops principal component models for each training set category. Its main objective is the reliable classification of new samples. When a prediction is made with the SIMCA technique, new samples insufficiently close to the PC space of a class are considered nonmembers. Furthermore, the technique requires that each training sample be pre-assigned to one of Q different categories, where Q is typically greater than one. It provides three possible outcome predictions: the sample fits only one pre-defined category, the sample does not fit any of the pre-defined categories, and the sample fits into more than one pre-defined category [67].

3.1.3 Computers, software, compounds, and molecular descriptors

For the present chapter, we performed molecular calculations on an AMD PHENOM 955 X4 2.2 GHz processor with 4 Gb of RAM with the Gaussian 98 program package [69]. The MEP was computed from the electronic density, and the maps were displayed using the MOLEKEL software [70], while the PR models were carried out on a PC Pentium machine with the Pirouette program [71].

Figure 1 shows the 2D structure of the 5-nitrofuran-2-aldoxim molecule [72] used in the selection of method/basis set (see Section 3.1.3.1). In **Figures 2** and **3** the 2D structures of the nitrofuran compounds from the training [73–75] and prediction sets are displayed, respectively. In this work, the nitrofuran molecules were defined as more active against *T. cruzi*, when in vitro *growth rate inhibition (GR) T. cruzi* ≥ 75, and as less active when in vitro *growth rate inhibition T. cruzi* < 75.

In general, the structure–activity relationship shows that for the compounds **1–6**, the increase in the carbon chain improves the activity against *T. cruzi*. The comparison between compounds **3** and **2** evidences increased activity by the substitution of the N atom by O. We can also notice that increasing the number of unsaturations and returning the nitrogen to the chain will lead to a decrease in biological activity (**7**, **8**). Still in relation to compound **1**, increasing the unsaturations, returning the atom of O, and increasing the carbon chain length (**9–12**) substantially increase the activity against *T. cruzi*. On the other hand, in compounds **13** and **14**, returning to an unsaturation in the main chain and introducing electron-withdrawing groups and more electronegative atoms, there is a decrease in chagasic activity. This evidence can also be verified for compounds **16**, **17**, **19–22**.

The molecular descriptors were obtained for the most stable conformation of each compound. These descriptors were computed to give information about the influence of electronic, steric, hydrophilic, and hydrophobic features on the anti-trypanosomal activity of the studied nitrofurans. The atomic charges in this work were derived from the electrostatic potential obtained with HF/6-31G method/basis

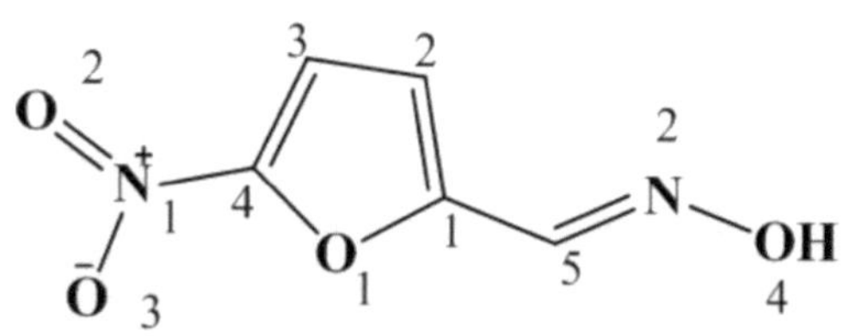

Figure 1.
2D molecular structure for 5-nitrofuran-2-aldoxime.

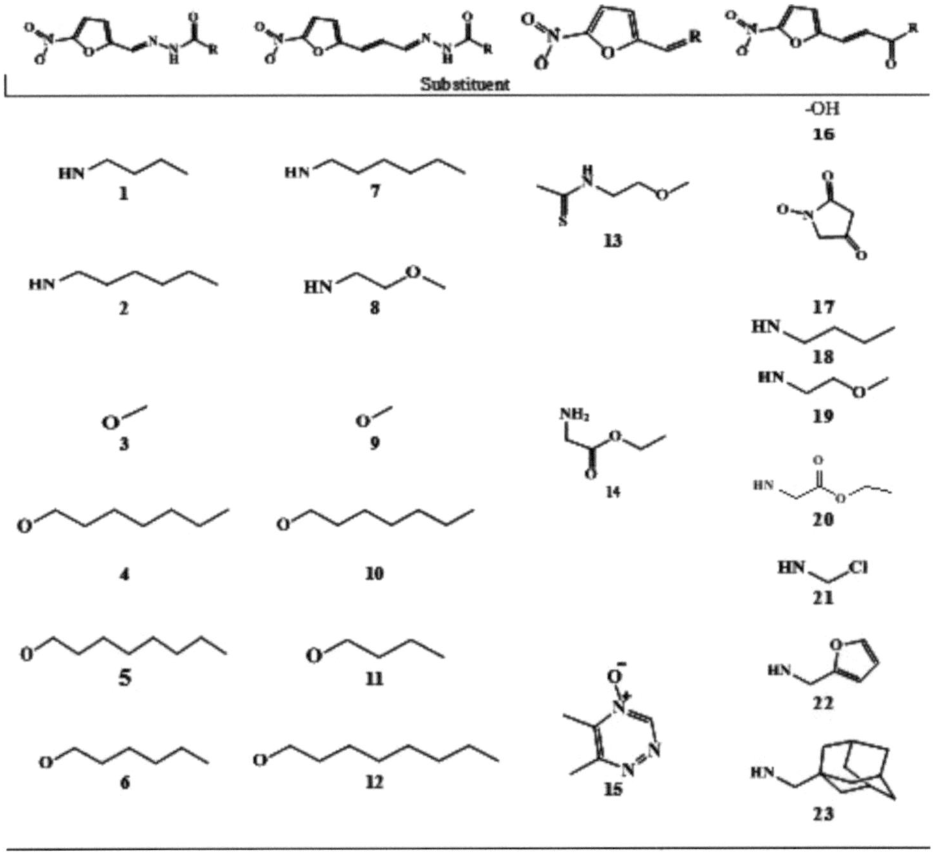

Figure 2.
2D molecular structure for nitrofurans (training set).

set as implemented in the Gaussian program package. The electrostatic potential is obtained through the calculation of a set of punctual atomic charges so that it represents the possible best quantum molecular electrostatic potential for a set of points defined around the molecule [76, 77]. The charges derived from electrostatic potential present the advantage of being, in general, physically more satisfactory than the charges of Mülliken [78], especially with regard to biological activity.

The quantum–chemical descriptors employed and obtained with the Gaussian 98 program package [69] were total energy of molecules (TE), highest occupied molecular orbital (HOMO) energy, one level below to highest occupied molecular orbital (HOMO–1) energy; lowest unoccupied molecular orbital (LUMO) energy, one level about lowest unoccupied molecular orbital (LUMO+1) energy, HOMO energy–LUMO energy (gap energy), total dipole moment (μ), Mulliken's electronegativity (χ), atomic charges on the Nth atom (QN), molecular hardness (HD), and molecular softness (MS).

The physicochemical descriptors obtained with ChemPlus module [79] were total surface area (TSA), molecular volume (VOL), molecular refractivity (MR), and molecule hydration energy (MHE).

Molecular holistic (MH) descriptors were included with the purpose of representing different sources of chemical information in terms of molecular size, symmetry, and distribution of atoms in molecules. Also, we include topologic indices, connectivity indices, geometric descriptors, 3D-MoRSE descriptors, and Moriguchi octanol–water partition coefficient (MlogP). These descriptors were calculated with the Dragon software [80].

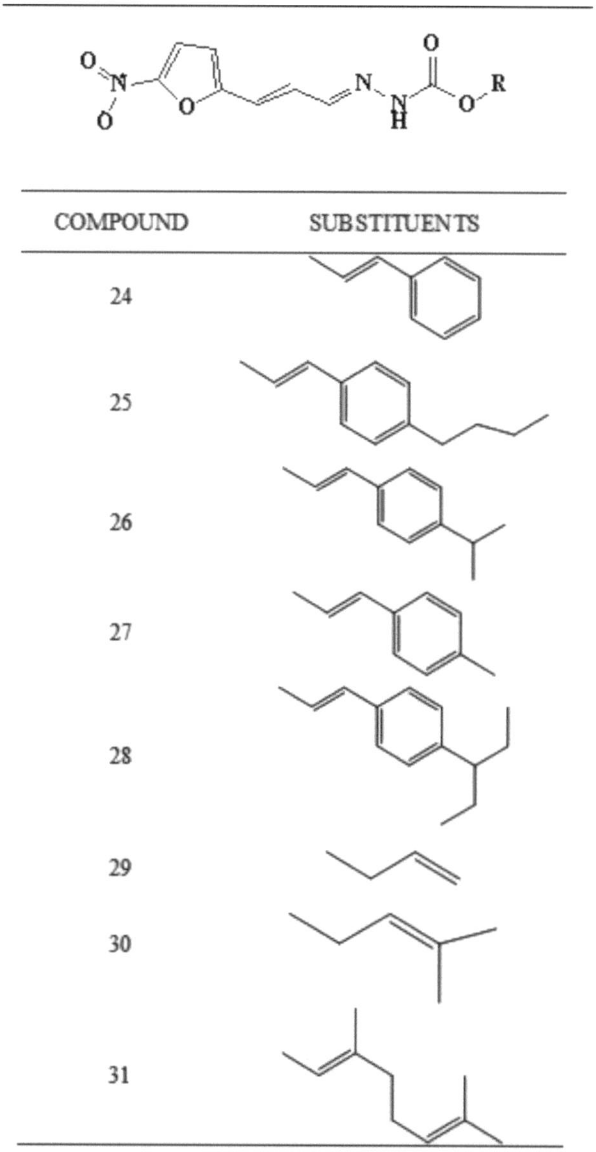

Figure 3.
2D molecular structures for nitrofurans for the prediction set.

3.1.3.1 Theoretical approach and basis set used in the molecular calculations

In the calculations with the nitrofuran compounds (**Figure 1**), quantum–chemical approaches were used [81–87]. We use Becke's three-parameter hybrid methods [81], the Lee-Yang-Parr (LYP) correlation functional [82], B3LYP and Becke's 1988 functional (BLYP) [83], Hartree-Fock (HF) method [84], Austin model 1 (AM1) method [85], Parametric Method Number 3 (PM3) [86], and standard basis sets [87] available in the Gaussian program package. In 5-nitrofuran-2-aldoxim, geometry optimization was carried out by B3LYP/6-21G, B3LYP/6-21G*, B3LYP/6-31G, B3LYP/6-31-G*, BLYP/6-21G, BLYP/6-21G*, BLYP/6-31G, BLYP/6-31G*, HF/6-21G, HF/6-21G*, HF/6-31G, and HF/6-31G* approaches [81–84] and basis sets [87] and AM1 and PM3 approaches [85, 86] . The calculations were performed to find the approach and basis set that would present the best compromise between computational time and accuracy of the information relative to the experimental data. The experimental structure of 5-nitrofuran-2-aldoxim molecule was retrieved from the Cambridge Structural Database CSD [72]. PCA and HCA techniques were used to compare the computed structures with different methods/basis sets of quantum chemistry with the experimental structure of 5-nitrofuran-2-aldoxim molecule to identify the appropriate method and the basis set for further calculations. The analyzes were carried out on an auto-scaled data matrix with dimension 26 × 5, where each row was associate 26 computed and 1 experimental geometry, and each column represented one of 5 geometrical parameters of the 5-nitrofuran-2-aldoxim molecule (bond lengths and bond angles). In order to compute all structures and perform calculations to obtain the molecular properties, the HF/6-31G method has selected (see Results and discussion section); the initial geometries of the nitrofurans (**Figures 2** and **3**) were built with the optimized geometry of the 5-nitrofuran-2-aldoxim molecule selected by PCA and HCA techniques. A conformational analysis for each compound was carried out with the MM^+ algorithm [79], and the lowest energy conformation was submitted to a conformational search with the Gaussian program.

3.2 Results and discussion

3.2.1 Quantum–chemical approach and basis set selection for the description of the geometries of nitrofurans

The advantage in using the PCA and HCA techniques in this step was that all structural information are considered simultaneously and it takes into account the correlations among them. **Table 1** shows the theoretical and experimental structural information (bond lengths and bond angles) of the geometry of the 5-nitrofuran-2-aldoxim molecule. It was used with the aim to select using PCA and HCA techniques, which quantum–chemical approach and basis set give results closest to the experimental data [72].

The first two principal components explain 86.02% of the original information as follows: PC1 = 58.01% and PC2 = 28.02%. The PC1 versus PC2 scores plot is shown in **Figure 4**, from which it can be seen that the methods are discriminated into two classes according to PC2. The semiempirical approaches (AM1 and PM3) are at the top of the graph, while the other theoretical (HF, BLYP, and B3LYP) approaches and experimental data are at the bottom. Moreover, it can be seen that the HF/6-31G approach/basis set is the closest to the experimental data, indicating that they should be used in the development of this work.

Also, to investigate the most appropriate approach and basis set for further calculations, we used HCA. **Figure 5** shows the dendrogram obtained with complete linkage method; from this figure, we conclude that the theoretical approaches are distributed in a similar way as in PCA, i.e., HCA confirmed the PCA results. Moreover, we can observe that the HF/6-31G approach/basis set is closer to the experimental data therefore being the most suitable to carry out this work.

3.2.2 MEP maps for compounds of the training set

Figure 6 shows the MEP maps for the nitrofurans in the training set. The analysis of these maps reveals that the most active compounds, in general, have the following characteristics:

(i) Compounds with an unsaturation and presenting O atom neighboring the carbonyl in the carbonic chain present greater electron density in the proximities of the furan ring with the decrease of the chain size. In these compounds (**4, 5,** and **6**), MEP

Approaches/basis set															
Geometric parameters	**B3LYP/6-21G**	**B3LYP/6-21G***	**B3LYP/6-31G**	**B3LYP/6-31G***	**BLYP/6-21G**	**BLYP/6-21G***	**BLYP/6-31G**	**BLYP/6-31G***	**HF/6-21G**	**HF/6-21G***	**HF/6-31G**	**HF/6-31G***	**AM1**	**PM3**	**Exp [72]**
Bond length (Å)															
C_2C_3	1.42	1.42	1.42	1.42	1.43	1.47	1.43	1.42	1.43	1.43	1.43	1.43	1.43	1.43	1.41
C_4C_5	1.36	1.36	1.37	1.37	1.38	1.38	1.39	1.38	1.34	1.39	1.34	1.34	1.40	1.39	1.34
C_1C_2	1.38	1.38	1.38	1.38	1.39	1.39	1.40	1.39	1.35	1.35	1.35	1.35	1.33	1.38	1.36
C_1O_1	1.40	1.37	1.39	1.36	1.42	1.39	1.42	1.38	1.37	1.39	1.37	1.33	1.34	1.37	1.37
C_4O_1	1.38	1.35	1.38	1.35	1.41	1.37	1.40	1.37	1.36	1.37	1.35	1.33	1.40	1.38	1.35
C_4N_1	1.41	1.43	1.41	1.43	1.43	1.44	1.43	1.49	1.40	1.43	1.40	1.42	1.45	1.48	1.42
N_1O_2	1.41	1.43	1.41	1.43	1.43	1.44	1.43	1.48	1.40	1.43	1.41	1.42	1.46	1.47	1.42
N_1O_3	1.28	1.23	1.26	1.23	1.31	1.25	1.29	1.25	1.24	1.19	1.22	1.19	1.19	1.21	1.22
C_1C_5	1;29	1.23	1.27	1.23	1.32	1.26	1.30	1.26	1.26	1.20	1.23	1.20	1.20	1.22	1.22
C_5N_2	1.43	1.44	1.43	1.44	1.44	1.45	1.44	1.45	1.45	1.46	1.45	1.46	1.45	1.45	1.45
N_2O_4	1.29	1.28	1.29	1.28	1.32	1.31	1.31	1.30	1.26	1.25	1.26	1.25	1.31	1.29	1.27
O_4H_1	1.47	1.40	1.44	1.39	1.50	1.42	1.47	1.41	1.44	1.37	1.40	1.36	1.31	1.39	1.38
Bond angle (°)															
$C_1O_1C_4$	105.3	105.6	106.0	106.1	104.7	105.3	105.5	105.8	105.4	106.3	105.8	106.9	105.3	106.3	104.5
$O_1C_1C_2$	109.5	110.2	109.2	110.1	109.5	110.0	109.2	109.9	109.1	109.6	110.7	109.4	105.2	106.0	104.8
$O_1C_1C_5$	119.3	118.7	119.8	119.5	119.2	118.8	119.7	119.6	119.5	119.4	118.5	119.8	110.6	110.7	110.2
$C_5C_1C_2$	131.2	131.0	130.9	130.4	131.3	131.1	131.1	130.5	131.3	130.9	130.6	130.9	119.5	120.4	114.1
$C_5N_2O_2$	121.2	120.6	121.8	121.3	121.2	120.7	121.9	121.4	122.0	120.9	120.1	121.7	129.7	128.8	135.6
$C_1O_1C_4$	107.9	110.0	109.5	110.6	107.1	109.2	108.5	109.8	108.8	109.6	111.2	111.4	122.8	120.4	127.8
$N_2O_4H_1$	100.7	100.8	103.6	102.7	99.3	99.8	102.0	101.6	102.4	103.7	102.1	106.9	115.2	116.7	112.2

$C_1C_2C_3$	107.5	106.6	107.5	106.6	107.7	106.8	107.7	106.9	107.8	106.9	106.0	106.9	104.2	101.6	106
$O_1C_4C_3$	111.5	112.3	111.2	112.0	111.5	112.2	111.2	111.9	111.1	111.4	112.8	111.1	105.9	106.0	105.1
$C_3C_4N_1$	130.2	129.9	130.6	130.2	130.1	130.0	130.9	130.5	130.9	130.3	129.5	130.4	111.1	110.6	113.2
$O_1C_4N_1$	118.4	117.8	118.1	117.6	118.3	117.7	117.9	117.5	117.8	118.2	117.5	118.4	131.4	131.4	129.8
$C_4N_1O_2$	115.0	114.9	115.8	115.6	114.6	114.6	115.9	115.5	115.7	115.4	115.0	116.0	117.4	117.8	116.9
$C_4N_1O_3$	117.7	117.2	118.9	118.1	117.6	117.2	115.6	118.2	118.9	118.1	117.2	119.2	117.3	117.5	116.3
$O_2N_1O_3$	127.3	127.9	125.3	126.2	127.7	128.1	118.8	126.2	125.2	126.4	127.6	126.4	119.6	120.1	118.8

**Refers to the base sets cited in the corresponding references.*

Table 1.
Experimental and theoretical structural parameter of the 5-nirofuran-2-aldoxime.

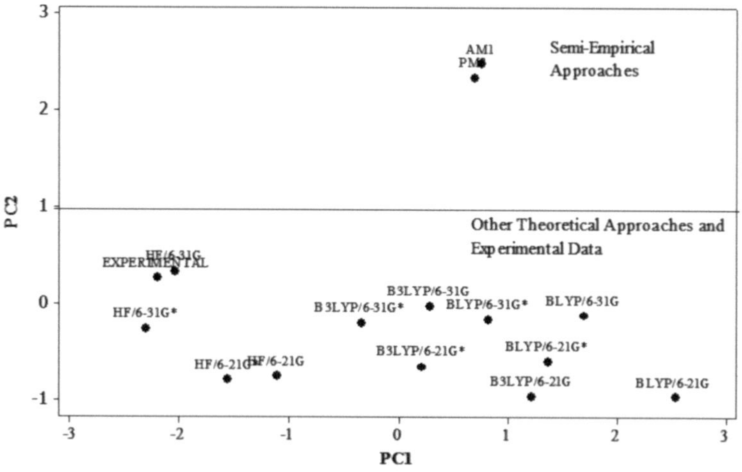

Figure 4.
Score plots of the two first PCs, PC1 and PC2, for the separation of the approaches basis sets into classes: semiempirical and semiempirical not.

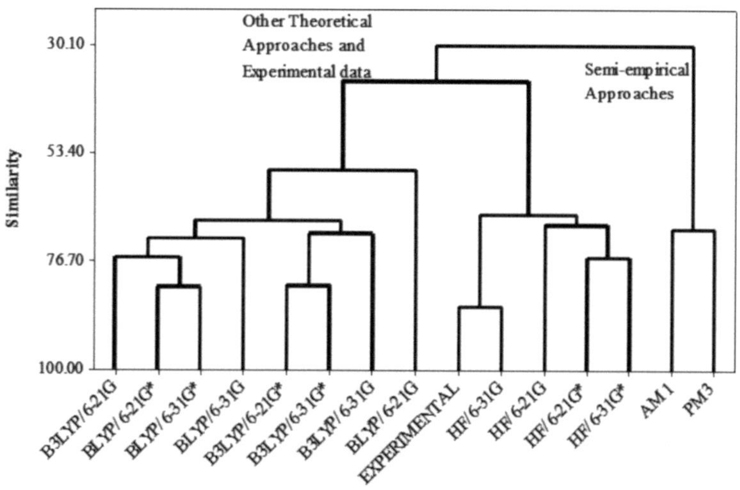

Figure 5.
Dendrogram obtained with HCA technique for the separation of the approach basis set into two classes: semiempirical and semiempirical not.

maps show negative regions ranging from −82.99 to −4.87 kcal/mol. In the most active compound (**6**), as can be seen, the most negative values are in the nitro group, the O atom of the furan ring and the O atoms of the ester group (red and yellow). Also, the MEP maps of these compounds exhibit positive regions between the +4.54 and + 76.96 kcal/mol values (green and blue). Compounds with double unsaturation, containing N atom next to the carbonyl, raise the electronic density with the increase of the carbonic chain. In the most active compound (7), the MEP map shows a region of negative values between −77.74 and − 1.31 kcal/mol, with the electron density concentrating mainly on the atoms of the nitro group, on the O atom of the furanic ring and on the N and O atoms of the amide group (red and yellow). According to the MEP map, these compounds present positive MEP between +5.64 and 61.21 kcal/mol (green and blue).

(ii) Compounds with double unsaturation, containing O atom neighboring the carbonyl, raising the carbon chain, increase the electron density in the atoms of the

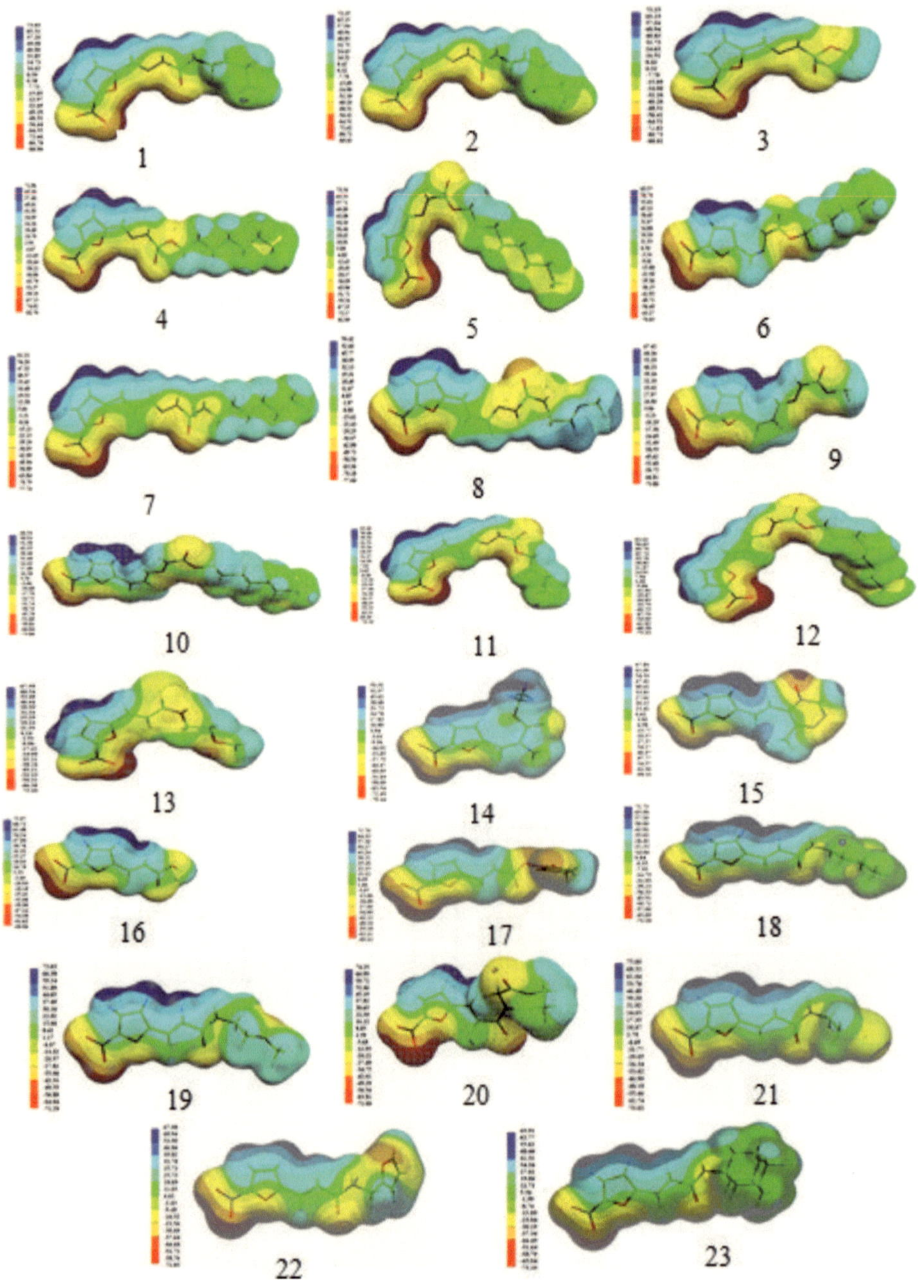

Figure 6.
MEP maps (kcal/mol) for nitrofurans (training set).

nitro group, extending through the O atom of the furan ring to the O atoms of the ester group following the unsaturated chain. In these compounds (**10–12**), the MEP maps exhibit more negative values between −76.18 and − 6.36 kcal/mol (red and yellow). They exhibit positive MEP in the range of +0.63 to 67.42 kcal/mol (green and blue)

(iii) Compound with an unsaturation, N atom neighboring the carbonyl in the carbonic chain and bulky substituents, has higher electron density in the vicinity of the furan ring and in the N and O atoms of the amide group. In this compound

(**23**), the MEP map shows a negative region (red and yellow) between −73.10 and − 1.59 kcal/mol on the mentioned atoms and positive region between +5.56 and 69.91 kcal/mol (green and blue). The electron density around the nitro group, the O atom of the furan ring, and other atoms may induce the nitrofurans to show antitrypanosomal activity, suggesting the complexation in those regions with the active site of the receptor in a biological recognition process.

From the above discussion, as a rule, to plan more active nitrofurans, we can assume we resort to one of the basic structures of the most active compounds and introduce groups of atoms or substituents electron donors enhancing the key structural features that are necessary for their activities.

3.2.3 Chemometric modeling

To perform the chemometric modeling, all variables were auto-scaled as preprocessing so that they could be standardized and so they could have the same importance regarding the scale. Furthermore, given a large quantity of multivariate data available, it was necessary to reduce the number of variables. Thus if any two descriptors had a high Pearson correlation coefficient ($r > 0.8$), one of the two was excluded from the matrix at random, since theoretically they describe the same property [88]; they also have a high correlation with antitrypanosomal activity, and only one of them is enough to be used as independent variable in a predictive model.

3.2.3.1 PCA model

Four molecular descriptors were selected for PCA model. The molecular descriptors (QN1, gap energy, Mor05u, and MlogP), in vitro *T. cruzi* growth inhibition (experimental data), and activity and correlation matrix including all data for 23 nitrofurans can be seen in **Table 2**. The correlation between descriptors is less than 0.786. The first three principal components (PCs) describing 96.48 of the original information for the 23 are as follows: 45.70, 30.91, and 19.87%. PC1-PC2 scores for the samples are shown in **Figure 7**. From this figure, we can see that the nitrofurans are distributed into two distinct regions in PC1. The more active compounds are on the left side (**4–7, 10–12, 18,** and **23**) and the less active on the right side (**1–3, 8, 9, 13–17,** and **19–22**). According to **Figure 8**, the MlogP descriptor is responsible for displaying more active compounds on the left side, while the gap energy, QN1, and Mor05u descriptors displayed fewer active compounds for the right side from this figure.

Table 3 shows the loading vectors for PC1, PC2, and PC3. According to this table, PC1 can be expressed through the following equation:

$$PC1 = 0.20\ (QN1) + 0.06\ (Gap\ energy) + 0.71\ (Mor05u) - 68\ (MlogP). \quad (2)$$

From this equation, more active nitrofurans, in general, can be obtained when we have lower values for the QN1 combined with lower values for Gap energy and Mor05u and higher values for MlogP.

3.2.3.2 HCA model

The results of the HCA model are displayed in the dendrogram in **Figure 9** and are similar to those of PCA model. The nitrofurans are fairly well grouped according to their activity. From this figure, the two clusters (+ and −) mirror the same two classes displayed by PCA model (**Figure 7**).

Nitrofurans	QN1	Gap energy (kcal/mol)	Mor05u	MlogP	% in vitro *T. cruzi* growth inhibition[a,b]	Activity[c]
1−	0.201	220.9	−3.966	1.135	30	LA
2−	0.201	220.9	−2.938	1.708	20	LA
3−	0.165	220.9	−2.723	0.181	32	LA
4+	0.165	226.5	−6.869	1.980	92.7	MA
5+	0.165	225.3	−7.439	3.155	83.7	MA
6+	0.169	229.7	−0.016	1.708	96.2	MA
7+	0.164	208.3	−7.439	1.889	81.9	MA
8−	0.164	205.2	−4.854	0.334	26.7	LA
9−	0.166	215.9	−3.292	0.478	58	LA
10+	0.166	215.9	−7.470	2.146	90	MA
11+	0.164	208.3	−5.674	1.354	87.4	MA
12+	0.164	208.3	−8.435	3.307	92.3	MA
13−	0.167	195.2	−4.338	0.751	12	LA
14−	0.161	203.3	−2.872	0.501	3	LA
15−	0.167	208.3	−4.217	0.411	30	LA
16−	0.167	225.3	−2.373	0.609	20	LA
17−	0.167	225.9	−4.054	1.063	6	LA
18+	0.167	225.3	−6.339	2.001	75	MA
19−	0.166	225.3	−4.145	0.398	31	LA
20−	0.167	226.5	−4.786	0.667	35	LA
21−	0.167	225.3	−3.398	1.157	23	LA
22−	0.166	218.4	−3.876	0.802	14	LA
23+	0.166	224.6	−6.314	3.014	90.5	MA
Gap energy	−0.171					
Mor05u	0.27	−0.006				
MlogP	0.026	−0.184	−0.785			

[a]*Inhibitor concentration of 5 μM.* [b]*Growth inhibition ≥ 75, more active (MA)*[c]*, and growth inhibition < 75, less active (LA)*[c]*.*

Table 2.
Values for the four most important descriptors which classify the studied nitrofuran compounds, in vitro T. cruzi growth inhibition (experimental data), activity, and correlation matrix.

3.2.3.3 KNN model

Table 4 shows the results for the KNN models obtained with the KNN technique and constructed with one (1NN) to four (4NN) nearest neighbors. To all models the percentage of correct information was 100%. We used the model 4NN because the greater the number of the nearest neighbors, the better the reliability of the KNN technique, and the same was used for validation of the training set from **Figure 2**.

3.2.3.4 SDA model

In the construction of the SDA model, the discrimination functions for groups more active and less active, respectively, are given below:

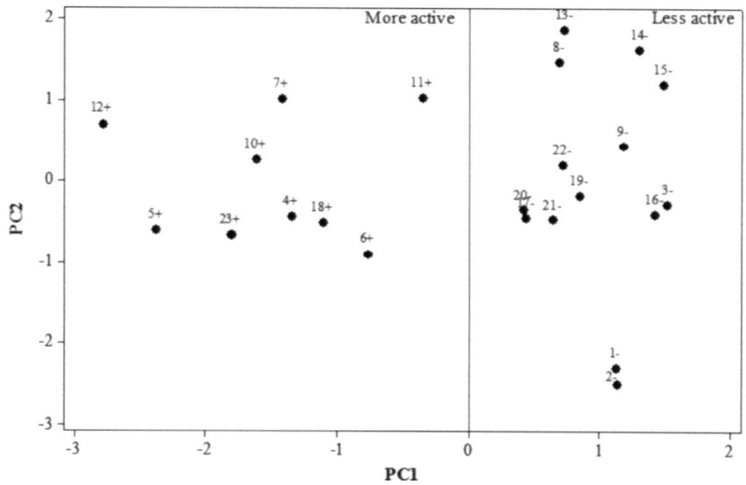

Figure 7.
Score plots of the two first PCs, PC1 and PC2, responsible for the separation of the 23 nitrofurans (training set) into two classes: (+) more active and (−) less active against T. cruzi.

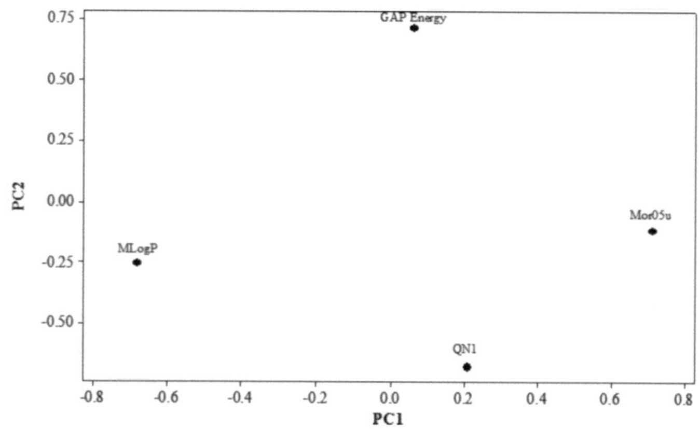

Figure 8.
Loading vector plots of the first PCs, PC1 and PC2, for four variables responsible for the separation of the 23 nitrofurans (training set) into two classes: (+) more active and (−) less active against T. cruzi.

Group MA (more active):

$$0.51(QN1) + 0.43Gap\ energy + 3.05Mor05u - 1.5MlogP - 0.62 \quad (3)$$

Group LA (less active):

$$-0.80QN1 - 0.67Gap\ energy - 4.75Mor05u + 2.34MlogP - 3.92 \quad (4)$$

Also, through the discrimination functions, Eqs. (3) and (4), and of the value of each descriptor for the nitrofurans, we obtain the classification matrix by using all compounds from the training set (**Table 5**). The classification error was 0.00% resulting in a satisfactory separation of more active and less active compounds. From SDA model, the allocation rule was derived when the activity against *T. cruzi* of new nitrofurans is investigated: (a) initially calculate, for the new compound, the value of the most important descriptors obtained in the construction of the SDA model, (b) put these auto-scaled values in the two discrimination functions

Variable	PC1	PC2	PC3
QN1	0.20	0.66	0.69
Gap energy	0.06	−0.70	0.70
Mor05u	0.71	0.11	−0.10
MlogP	−0.68	0.26	0.17

Table 3.
Variables matrix for the first three principal components.

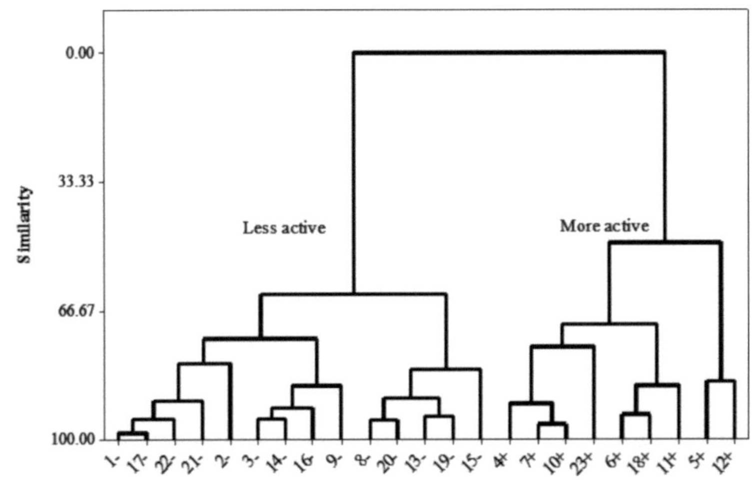

Figure 9.
Dendrogram obtained with HCA technique for the separation of the nitrofurans into two classes: (+) more active and (−) less active against T. cruzi.

Category	Number of compounds	Compounds incorrectly classified			
		1NN	2NN	3NN	4NN
Class:more active	9	0	0	0	0
Class: less active	14	0	0	0	0
Total	23	0	0	0	0
% Correct information		100	100	100	100

Table 4.
Classification obtained with the KKN technique.

performed in this work, and (c) check which discrimination function, Eq. (3) or Eq. (4), presents higher value. The new compound is more active if it is related to discrimination function of group more active and vice versa.

In order to check the reliability of the model, the "leave-one-out technique" was employed. One nitrofuran compound is excluded from the data set, and the remaining compounds are used in building the classification functions.

Subsequently, the removed analogue is classified according the generated classification functions. In the further step, the omitted compound is included, and a new nitrofuran is removed, and the procedure goes on until the last compound is removed. In **Table 6** the results obtained with the cross-validation model are summarized.

		True group	
Classification group or class	**Number of compounds**	**More active**	**Less active**
Group (Class): more active	9	9	0
Group (Class): less active	14	0	14
Total	23	9	14
% Correct information	—	100	100

Table 5.
Classification matrix obtained using SDA technique.

		True group	
Classification group or class	**Number of compounds**	**More active**	**Less active**
Group (class): more active	9	9	0
Group (class): less active	14	0	14
Total	23	9	14
% correct information	—	100	100

Table 6.
Classification matrix obtained by using SDA technique with cross-validation technique.

3.2.3.5 SIMCA model

The SIMCA model were built with the same descriptors as PCA, HCA, KNN, and SDA models and used two (2) PCs in the modeling of the two classes: more active nitrofurans (**4–7**, **10–12**, **18,** and **23**) and less active (**1–3**, **8**, **9**, **13–17**, and **19–22**) nitrofurans. In **Table 7**, the obtained results for the SIMCA model are shown. In this case, the information percentage was also 100%. According to the PCA, HCA, KNN, SDA, and SIMCA models, we can also notice that the QN1, gap energy, Mor05u, and MlogP descriptors are key properties for explaining the anti-*T. cruzi* activity of the nitrofurans training set (**Figure 2**).

As QN1, gap energy, Mor05u, and MlogP properties were selected in the chemometric modeling as the most important characteristics to describe the antitrypanosomal activity, some considerations about them may be relevant to the understanding of the behavior of more active nitrofurans. According to classical chemical theory, chemical interactions can be classified in two categories: electrostatic (polar) or orbital (covalent). Electrical charges in the molecule are indubitably the impelling cause of electrostatic interactions. It has been demonstrated that local electron densities or charges are important in many chemical reactions, physicochemical properties, and ligand–receptor interactions [89, 90]. Thus, charge-based parameters have been widely employed as chemical reactivity indices or as measures of weak intermolecular interactions. Many quantum–chemical descriptors are derived from the partial charge distribution in a molecule or from the electron densities on particular atoms [91]. From **Table 2**, we can observe that, in general, QN1 for more active analogues must present lower values than the less active ones. This is an indication that biological processes can occur through electrostatic interactions between the more active nitrofurans and an eventual biological receptor.

Gap energy is an important stability index. A large gap energy implies high stability for the molecule in the sense of its lower reactivity in chemical reactions.

Category	Number of compounds	Correct classification
Class: more active	9	9
Class: less active	14	14
TOTAL	23	
% correct information		100

Table 7.
Classification obtained by using SIMCA technique.

It is an approximation of the lowest excitation energy of the molecule and can be used for the definition of absolute and activation hardness [89, 90]. In **Table 2**, we can observe that, in general, the more active nitrofurans present lower gap energy than the less active ones. This indicates that the more active nitrofurans have a great probability of interacting with the biological receptor through a charge transfer mechanism.

Mor05u is a 3D-MoRSE descriptor based on the idea of obtaining information from 3D atomic coordinates through the transformed used in electrons diffraction studies [91] and is strictly related to the stereochemistry of the compounds [92]. According to **Table 2**, the more active nitrofurans present lower values of Mor5u. This may be, in general, an indication of the importance of the stereochemical properties of the more active nitrofurans in a possible mechanism of action of its own.

MlogP is an important hydrophobic descriptor in diverse biochemical, pharmacological, and toxicological processes involved in drug absorption [93]. As identified in **Table 2**, the more active reported nitrofurans exhibit the higher MlogP values. This is an indication that in processes involving nitrofurans and a biological receptor, hydrophobic interactions may be important in the mechanism of action of these compounds.

Knowing the performance of the RP models constructed for the 23 studied nitrofurans, we decided to apply them to a series of eight compounds (**Figure 3**) designed to maintain the key structural features that are necessary for their biological activities evidenced by the MEP maps of the compounds of the training set. The basic nucleus of these compounds corresponds to that of the most active nitrofurans with double unsaturation, containing vicinal O atom to carbonyl (see compounds **10–12**). The eight molecules proposed for the study of prediction of activity were drawn with the help of one of the collaborators of this work, who belong to the research group in organic chemistry of the Federal University of Pará, Brazil, and the most promising syntheses are in progress. In the future, antitrypanosomal tests with the most promising nitrofurans can be used to validate our RP models.

The results obtained of the application of the PR models (PCA, HCA, KNN, SDA, and SIMCA) and the descriptors for the compounds of the prediction set are summarized in **Tables 8** and **9**, respectively. In **Table 8**, the compounds **25** and **30** were predicted as more active against *T. cruzi* with the five models. Only the KNN model predicted compound **26** as the most active. Meanwhile, only the PCA and HCA models predicted compound **31** as the most active. On the other hand, all models, except the SDA model, predicted compounds **24**, **27**, and **28** as the most active. In turn, the SIMCA model did not classify compounds **29** and **31** into any of the two classes. Thus, we can consider nitrofurans **25** and **30** as potentially more active in a future test against *T. cruzi*. For the values reported for compounds **25** and **30** (**Table 9**), it can be shown that in order to design more active nitrofurans we must combine smaller values for the descriptors QN1, gap energy, and Mor05u with higher value for the descriptor MlogP.

Nitrofuran	PCA model	HCA model	KNN model	SDA model	SIMCA model
24	MA	MA	MA	LA	MA
25	MA	MA	MA	MA	MA
26	LA	LA	MA	LA	LA
27	MA	MA	MA	LA	MA
28	MA	MA	MA	LA	MA
29	MA	MA	MA	MA	0
30	MA	MA	MA	MA	MA
31	MA	MA	LA	LA	0

Table 8.
Results of application of chemometric models for the nitrofurans of the prediction set.

Nitrofuran	QN1	Gap energy (kcal/mol)	Mor05u	MLogP
24	0.165	205.2	−6.352	3.155
25	0.165	203.3	−7.332	2.250
26	0.165	204.6	−5.835	1.146
27	0.169	203.9	−6.164	2.508
28	0.166	203.9	−7.146	1.875
29	0.164	229.7	−8.201	3.854
30	0.164	229.7	−6.421	3.373
31	0.164	223.4	−5.525	2.167

Table 9.
Values for descriptors for the prediction set.

3.2.4 MEP maps for compounds of the prediction set

Figure 10 shows the MEP maps for the most active nitrofurans in the validation set (**25** and **30**). Also, in these compounds, as can be seen, raising the carbon chain increases the electron density in the atoms of the nitro group, extending through the O of the furan ring to the O atoms of the ester group accompanying the unsaturated chain. In these compounds, the MEP maps show more negative values between −74.27 and − 1.76 kcal/mol (red and yellow). They exhibit positive MEP in the range + 4.84 to +57.58 kcal/mol (green and blue).

The negative MEP region of compounds **25** and **30**, similar to the more active compounds in the training set, is susceptible to attack in a biological recognition process.

3.3 Concluding remarks

MEP and chemometric techniques in the last decades have become efficient tools in the study of the structure–activity relationships of bioactive molecules. The use of such tools has occurred through the inherent principles of each or combining their potentials to more efficiently unravel information about the structure–activity relationships of pharmacologically interesting compounds. This chapter is circumscribed in this second possibility. MEP maps were constructed for 23 nitrofurans with activity against *T. cruzi* reported in the literature. The key structural features

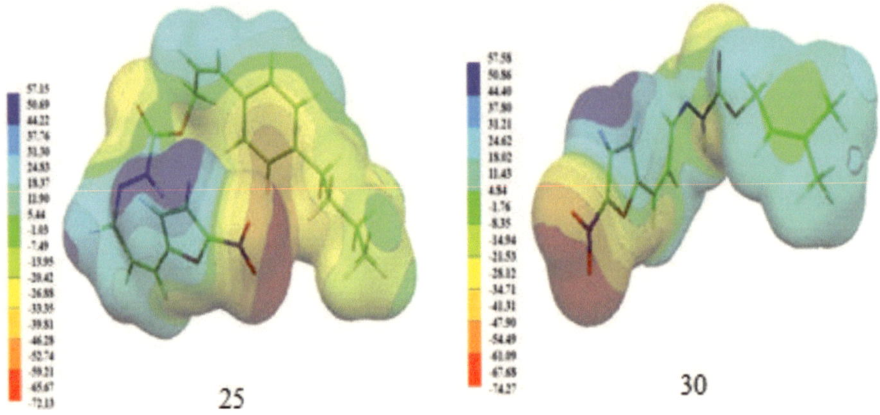

Figure 10.
MEP maps (kcal/mol) for most promising nitrofurans in the prediction set against T. cruzi.

required for antitrypanosomal activity, along with chemical intuition, allowed the introduction of substituents in one of the most active nitrofurans in the training set to obtain eight new derivatives.

PR models (PCA, HCA, KNN, SDA, and SIMCA) were constructed and demonstrated that 23 nitrofurans can be classified into two classes or groups: more active and less active according to their degrees of activity against *T. cruzi*. The properties QN1, gap energy, Mor05u, and MlogP are responsible for the classification into more active and less active studied nitrofurans. It is interesting to notice that these properties represent three distinct classes of interactions between the nitrofurans and the biological receptor: electronic (QN1 and gap energy), steric (Mor05u), and hydrophobic (MlogP). Here it is important to mention that Paulino et al., studying the influence of molecular parameters on the activity of 5-nitrofurans against *T. cruzi,* reported the importance of electronic properties and molecular hydrophobicity as well as the variation of the nitrofurans electronic structure to explain the greater activity of these compounds as inhibitors of the growth of this protozoan [94].

The results of the application of PR models on the validation set evidenced two nitrofurans (**25** and **30**) as more promising for synthesis and biological assays, which in the future can be used to validate our PR models.

Acknowledgements

We gratefully acknowledge the financial support of the Brazilian agencies: Conselho Nacional de Desenvolvimento Científico e Tecnológico and Coordenação de Aperfeiçoamento de Pessoal de Nível Superior. The authors would like to thank the Virtual Computational Chemistry Laboratory (VCCLAB–Munich) and the Swiss Center for Scientific Computing for the use of the DRAGON and MOLEKEL software, respectively. We employed computing facilities at the Laboratório de Química Teórica e Computacional (LQTC)–Universidade Federal do Pará.

Author details

Marcos Antônio B. dos Santos[1], Luã Felipe S. de Oliveira[2],
Antônio Florêncio de Figueiredo[3], Fábio dos Santos Gil[2],
Márcio de Souza Farias[2], Heriberto Rodrigues Bitencourt[4], José Ribamar B. Lobato[2],
Raimundo Dirceu de P. Farreira[2], Sady Salomão da S. Alves[3],
Edilson Luiz C. de Aquino[2] and José Ciríaco-Pinheiro[2]*

1 University of the State of Pará, Pará, Brazil

2 Computational and Theoretical Chemistry Laboratory, Federal University of Pará, Pará, Brazil

3 Federal Institute of Education, Science and Technology, Pará, Brazil

4 Group of Organic Chemistry, Federal University of Pará, Pará, Brazil

*Address all correspondence to: ciriaco@ufpa.br

References

[1] Bonnacorsi RR, Scrocco E, Tomasi J. Molecular SCF calculations for the ground state of some three-membered ring molecules: (CH2)3, (CH2)2NH, (CH2)2NH2+, (CH2)2O, (CH2)2S, (CH)2CH2, and N2CH2. The Journal of Chemical Physics. 1970;**52**:5270-5284. DOI: 10.1063/1.1672775

[2] Scrocco E, Tomasi J. The electrostatic molecular potential as a tool for the interpretation of molecular properties, in: New concepts II. Topics in Current Chemistry. 1973;**42**:95-170. DOI: 10.1007/3-540-06399-4

[3] Politzer P, Truhlar G, editors. Chemical Applications of Atomic and Molecular Electrostatic Potentials. New York: Plenum Press; 1981. ISSN: 978-4757-9634-6

[4] Rangel NL, Seminario JM. Molecular electrostatic potential devices on graphite and silicon surfaces. The Journal of Physical Chemistry A. 2006;**110**:12298-12302. DOI: 10.1021/jp064766i

[5] Müller JJ, Lapko A, Ruckpaul K, Heinemann U. Modeling of electrostatic recognition processes in the mammalian mitochondrial steroid hydroxylase system. Biophysical Chemistry. 2003;**100**:281-292. DOI: 10.1016/S0301-4622(02)00286-7

[6] Kotsikorou E, Sharir H, Shore DM, Hurst DP, Lynch DL, Madrigal KE, et al. Identification of the GPR55 antagonist binding site using a novel set of high-potency GPR55 selective ligands. Biochemistry. 2013;**52**:9456-9469. DOI: 10.1021/bi4008885

[7] Ford KA. Role of electrostatic potential in the in silico prediction of molecular bioactivation and mutagenesis. Molecular Pharmaceutics. 2013;**10**:1171-1182. DOI: 10.1021/mp3004385

[8] Politzer P, Murray JS. Quantitative analyses of molecular surface electrostatic potentials in relation to hydrogen bonding and co-crystallization. Crystal Growth & Design. 2015;**15**:3767-3774. DOI: 10.1021/acs.cgd.5b00419

[9] Lande DN, Gejji SP. Cooperative hydrogen bonding, molecular electrostatic potentials, and spectral characteristics of partial thia-substituted calix [4] arene macrocycles. The Journal of Physical Chemistry A. 2016;**120**:7385-7397. DOI: 10.1021/acs.jpca.6b07568

[10] Anjali BA, Sayyed FB, Suresh CH. Correlation and prediction of redox potentials of hydrogen evolution mononuclear cobalt catalysts via molecular electrostatic potential: A DFT study. The Journal of Physical Chemistry A. 2016;**120**:1112-1119. DOI: 10.1021/acs.jpca.5b11543

[11] Mehmood A, Jones SI, Tao P, Janesko BJ. An orbital-overlap complement to ligand and binding site electrostatic potential maps. Journal of Chemical Information and Modeling. 2018;**58**:1836-1846. DOI: 10.1021/acs.jcim.8b00370

[12] Liu L, Miao L, Li L, Li F, Lu Y, Shang Z, et al. Molecular electrostatic potential: A new tool to predict the lithiation process of organic battery materials. The Journal of Physical Chemistry Letter. 2018;**9**:3573-3579. DOI: 10.1021/acs.jpclett.8b01123

[13] Scilabra P, Murray JS, Terraneo G, Resnati G. Chalcogen bonds in crystals of bis(o-anilinium)diselenide salts. Crystal Growth & Design. 2019;**19**:1149-1154. DOI: 10.1021/acs.cgd.8b01634

[14] Pramanik S, Dey T, Mukherjee AK. Five benzoic acid derivatives: Crystallographic study using X-ray powder diffraction, electronic structure and molecular electrostatic potential calculation. Journal of Molecular Structure. 2019;**1175**:185-194. DOI: 10.1016/j.molstruc.2018.07.090

[15] Salluma LO, Vaza WF, Borgesa NM, Campos CEM, Bartoluzzib AJ, Francoc CHJ, et al. Synthesis, conformational analysis and molecular docking studies on three novel dihydropyrimidine derivatives. Journal of Molecular Structure. 2019;**1192**:274-287. DOI: 10.1016/j.molstruc.2019.04.100

[16] Rzesikowska K, Krawczuk A, Kalinowska-Tluscik J. Electrostatic potential and non-covalent interactions analysis for the design of selective 5-HT7ligands. Journal of Molecular Graphics and Modelling. 2019;**91**:130-139. DOI: 10.1016/j.jmgm.2019.06.007

[17] Aray Y. Nature of the active sites of molybdenum-based catalysts and their interaction with sulfur- and nitrogen-containing molecules using the quantum theory of atoms in molecules and the molecular electrostatic potential. The Journal of Physical Chemistry C. 2019;**123**:14421-14431. In press. DOI: 10.1021/acs.jpcc.9b01951

[18] Cruz JC, Hernández-Esparza R, Vázquez-Mayagoitia A, Vargas R, Garza J. Implementation of the molecular electrostatic potential over GPUs. Journal of Chemical Information and Modeling. 2019;**59**:3120-3127. in press. DOI: 10.1021/acs.jcim.8b00951

[19] Varmuza K. Pattern Recognition in Chemistry. 1980. Springer-Verlag, Berlin. DOI: 10.1002/bbpc.19810850930

[20] Johnson RA, Wichem DW. Applied Multivariate Statistical Analysis. New Jersey: Prentice-Hall; 1992. ISBN: 0-130-41146-9

[21] Brown SD, Sum ST, Despagne F, Lavine BK. Chemometrics. Analytical Chemistry. 1996;**68**:21R-61R. DOI: S0003-2700(96)00005-4

[22] Brown SD. The chemometrics revolution re-examined. Journal of Chemometrics. 2017;**31**:e2856. DOI: 10.1002/cem.2856

[23] Bernardinelli G, Jefford CW, Maric D, Thomson C, Weber J. Computational studies of the structures and properties of potential antimalarial compounds based on the 1,2,4-Trioxane ring structure. I. Artemisinin-like molecules. International Journal of Quantum Chemistry: Quantum Biology Symposium. 1994;**21**:113-131. DOI: 10.1002/qua.560520703

[24] Jefford CW, Grigorov M, Weber J, Lüthi HP, Troncher JMJ. Journal of Chemical Information and Computer Sciences. 2000;**40**:354-357 ISSN: 0095-2338

[25] Kubinyi H. QSAR: Hansch analysis and related approaches. In: Mannhold R, Krogsgaard-Larsen P, Timmerman H, editors. Methods and Principles in Medicinal Chemistry, Vol. 1. Weinheim: VHC; 1993. ISBN: 987-3527300358

[26] van de Waterbeemd H. Chemometric Methods in Molecular Design. New York: VHC; 2008. ISBN: 978-3-527-61544-5

[27] Gangwal RP, Damre MV, Sangamwar AT. Overview and recent advances in Qsar studies. Mercader AG, Duchwicz PR, Sivakumar PM, editors. CHEMOMETRICS: Applications and Research. QSAR in Medicinal Chemistry. Canada: Apple Academic Press; 2016. p. 1-32. ISBN: 978-1771-8811-35

[28] Kubinyi H, Folkers G, Martin YC, editors. 3DQSAR in Drug Design, Vols. 2 and 3. Dordrecht, The Netherlands: Kluwer; 1998. DOI: 978-0-7923-4791-0

[29] Cramer RD, Petterson DE, Brunce JD. Comparative molecular field analysis (CoMFA) 1. Effect of shape binding of steroids to carrier proteins. The Journal American of Chemical Society. 1988;**110**:5959-5967. DOI: 10.1021/ja00226a005

[30] Chhatbar DM, Chaube UJ, Vyas VK, Bhatt HF. CoMFA, CoMSIA, Topomer CoMFA, HQSAR, molecular docking and molecular dynamics simulations study of triazine morpholino derivatives as mTOR inhibitors for the treatment of breast cancer. Computational Biology and Chemistry. 2019;**80**:351-363. DOI: 10.1016/j.compbiolchem.2019.04.017

[31] Pourbasheer E, Aalizadeh R, Tabar SS, Ganjali MR, Norouzi P, Shadmanesh J. 2D and 3D quantitative structure–activity relationship study of hepatitis C virus NS5B polymerase inhibitors by comparative molecular field analysis and comparative molecular similarity indices analysis methods. Journal of Chemical Information and Modeling. 2014;**54**:2902-2914. DOI: 10.1021/ci500216c

[32] Cramer RD. Template CoMFA applied to 116 biological targets. Journal of Chemical Information and Modeling. 2014;**54**:2147-2156. DOI: 10.1021/ci500230a

[33] Cramer RD, Wendt B. Template CoMFA: The 3D-QSAR grail? Journal of Chemical Information and Modeling. 2014;**54**:660-671. DOI: 10.1021/ci400696v

[34] Dong H, Liu J, Liu X, Yu Y, Gao S. Molecular docking and QSAR analyses of aromatic heterocycle thiosemicarbazone analogues for finding novel tyrosinase inhibitors. Biooganic Chemistry. 2017;**75**:106-117. DOI: 10.1016/j.bioorg.2017.07.002

[35] Dong MH, Chen HF, Ren YJ, Shao FM. Molecular modeling studies, synthesis and biological evaluation of dabigatran analogues as thrombin inhibitors. Bioorganic & Medicinal Chemistry. 2016;**24**:73-84. DOI: 10.1016/j.bmc.2015.11.025

[36] Ding L, Wang ZZ, Sun XD, Yang J, Ma CY, Li W, et al. 3D-QSAR, molecular docking and molecular dynamics simulations study of 6-aryl-5-Cyano-Pyrimidine derivatives to explore the structure requirements of LSD1 inhibitors. Biooganic & Medicinal Chemistry Letters. 2017;**27**:3521-3528. DOI: 10.1016/j.bmcl.2017.05.065

[37] Pinheiro JC, Kiralj R, Ferreira MMC, Romero OAS. Artemisinin derivatives with antimalarial activity against plasmodium falciparum designed with the aid of quantum chemical and partial least squares methods. QSAR & Combinatorial Science. 2003;**22**:830-842. DOI: 10.1002/qsar.200330829

[38] Cardoso FJB, Figueiredo AF, Lobato MS, Miranda RM, Almeida RCO, Pinheiro JCP. A study on antimalarial artemisinin derivatives using MEP maps and multivariate QSAR. Journal of Molecular Moldeling. 2008;**14**:39-49. DOI: 10.1007/s00894-007-0249-9

[39] Ferreira JEV, Figueiredo AF, Barbosa JP, Cristino MGG, Macedo WJC, Silva OPP, et al. A study of new antimalarial artemisinins through molecular modeling and multivariate. Journal of the Serbian Chemical Society. 2010;**75**:1533-1548. DOI: 10.2298/JSC100126124F

[40] Figueiredo AF, Ferreira JEV, Barbosa JP, Macedo WJC, Cristino MGG, Lobato MS, et al. A computational study on antimalarial dispiro-1,2,4-trioxolanes. Journal of Computational and Theoretical Nanoscience. 2011;**8**:1-10. DOI: 10.1166/jctn.2011.1892

[41] Carvalho JRC, Ferreira JEV, Barbosa JP, Lobato MS, Meneses CCF,

Soeiro MM, et al. Computational modeling of artemisinins with antileishmanial activity. Journal of Computational and Theoretical Nanoscience. 2011;**8**:1-11. DOI: 10.166/jctn.2011.1943

[42] Barbosa JP, Ferreira JEV, Figueiredo AF, Almeida RCO, Silva OPP, Carvalho JRC, et al. Molecular modeling and chemometric study of anticancer derivatives of artemisinin. Journal of the Serbian Chemical Society. 2011;**76**:1263-1282. DOI: 10.2298/JSC111227111B

[43] Cristino MGG, Meneses CCF, Soeiro MM, Ferreira JEV, Figueiredo AF, Barbosa JP, et al. Computational modeling of antimalarial 10-substituted deoxoartemisinis. Journal of Theoretical and Computational Chemistry. 2012;**11**:241-263. DOI: 10.1142/S0219633612500162

[44] Politzer P, Laurence PR, Jayasuriya K. Molecular electrostatic potentials: An effective tool for the elucidation of biochemical phenomena. Environmental Health Perspectives. 1985;**61**:191-202. DOI: 10.1289/ehp.8561191

[45] Politzer P, Murray JS. The fundamental nature and role of the electrostatic potential in atoms and molecules. Theoretical Chemistry Accounts. 2002;**108**:134-149. DOI: 10.1007/s00214-002-0363-9

[46] Scrocco E, Tomasi J. Electronic molecular structure, reactivity and intermolecular forces: An euristic interpretation by means of electrostatic molecular potentials. Advances in Quantum Chemistry. 1979;**1978**(11):115-193. DOI: 10.1016/S0065-3276(08)60236-1

[47] Bishop CM. Pattern Recognition and Machine Learning. 2006. Springer, Singapore. ISBN: 978-0-387-31073-2

[48] Sebestyen GS. Decision-Making Processes in Pattern Recognition. New York: Academic Press; 1962

[49] Fu KS. Sequential Methods in Pattern Recognition and Machine Learning. New York: Academic Press; 1968. ASIN: B001QC5IZS

[50] Watanabe S. Methodologies of Pattern Recognition. New York: Academic Press; 1969. DOI: 978-1-4832-3093-1

[51] Mendel JM, Fu KS. Adaptive, Learning, and Pattern Recognition Systems; Theory and Applications, Vol. 66. 1st ed. New York: Academic Press; 1970. ISBN: 9780080955759

[52] González AG. Critical aspects of supervised Pattern recognition methods for interpreting compositional data. In: Varmuza K, editor. Chemometric in Applications Practical. Shanghai: In Tech; 2012. DOI: 978-953-51-0438-4

[53] Li Y, Wang S, Tian Q, Ding X. Feature representation for statistical-learning-based object detection: A review. Pattern Recognition. 2015;**48**:3542-3559. DOI: 10.1016/j.patcog.2015.04.018

[54] Jain AK. Data clustering: 50 years beyond K-means. Pattern Recognition Letters. 2010;**31**:651-666. DOI: 10.1016/j.patrec.2009.09.011

[55] Jurs PC, Kowalski BR, Isenhour TL. Investigation of combined patterns from diverse analytical data using computerized learning machines. Analytical Chemistry;**1969**:41, 1949-1953. DOI: 10.1021/ac50159a027

[56] Kowalski BR, Jurs PC, Isenhour TL. Computerized learning machines applied to chemical problem. Interpretation of infrared spectrometry. Analytical Chemistry. 1969;**41**:1945-1949. DOI: 10.1021/ac50159a026

[57] Kowalski BR, Reilly CA. Nuclear magnetic resonance spectral interpretation by Pattern recognition. The Journal Physical Chemistry. 1971;**75**:1402-1411. DOI: 10.1021/j100680a008

[58] Wangen LE, Isenhour TL. Semiquantitative analysis of mixed gamma-ray spectra by computerized learning machines. Analytical Chemistry. 1970;**42**:737-743. DOI: 10.1021/ac60289a005

[59] Sybrandt LB, Perone SP. Computerized learning machine applied to qualitative analysis of mixtures by stationary electrode polarography analytical chemistry. Analytical Chemistry. 1971;**43**:382-388. DOI: 10.1021/ac60322a009

[60] Isenhour TL, Jurs PC. Some chemical applications of machine intelligence. Analytical Chemistry. 1971;**43**:20. DOI: 10.1021/ac60304a037

[61] Kowalski BR, Brender CF. Pattern recognition. A powerful approach to interpreting chemical data. The Journal of American Chemical Society. 1972;**94**:5632-5639. DOI: 10.1021/ja00771a016

[62] Koskinen JR, Kowalski BR. Interactive pattern recognition in the chemical laboratory. Journal of Chemical Information and Computer Sciences. 1975;**15**:119-123. DOI: 10.1021/ci60002a010

[63] Kryger L. Interpretation of analytical chemical information by pattern recognition methods-A survey. Talanta. 1981;**28**:871-887. DOI: 10.1016/0039-9140(81)80223-8

[64] Danzer K, Singer R. Application of pattern recognition methods for the investigation of chemical homogeneity of solids. Mikrochimica Acta. 1985;**85**:219-226 ISSN: 0026-3672

[65] von Waterbeemd H, Tayar NE, Carrupt PA, Testa B. Pattern recognition study of QSAR substituent descriptors. Journal of Computer-Aided Molecular Design. 1989;**3**:111-132 ISSN: 0920-654X

[66] Laplante JP, Pemberton M, Hjelmfelt A, Ross J. Experiments on pattern recognition by chemical kinetics. The Journal Physical Chemistry. 1995;**99**:1063-10065. DOI: 10.1021/j100025a001

[67] Beebe KR, Pell RJ, Seasholtz MB. Chemometrics: A Practical Guide 1998. New York: Wiley & Sons; 1998. ISBN-10: 0471124516

[68] Mardia KV, Kent JT, Bibby JM. Multivariate Analysis. New York: Academic Press; 1979. ISBN: 9780124712522

[69] Frisch A, Frisch MJ. Gaussian 98 User 'S Reference, Revision a.7. Pittsburgh: Gaussian, Inc; 1998

[70] Flukiger P, Luth HP, Portmann S, Weber J. MOLEKEL 4.3. Manno, Switzerland: Swiss Center for Scientific Computing; 2000-2001

[71] Infometrix, Inc. Pirouette 3.01 2002, Woodinville

[72] Olszak TA, Peeters OM, Blaton NM, Ranter CJ. 5-Nitrofuran-2-aldoxime. Acta Crystallographica C. 1995;**51**:1304-1306. DOI: 10.1107/S0108270194008425

[73] Aguirre G, Cabrera E, Cerecetto H, Di Maio R, González M, Seoane G, et al. Design, synthesis and biological evaluation of new potent 5-nitrofuryl derivatives as anti-*Trypanosoma cruzi* agents. Studies of trypanothione binding site of trypanothione reductase as target for rational design. European Journal of Medicinal Chemistry. 2004;**39**(5):421-431. DOI: 10.1016/j.ejmech.2004.02.007

[74] Cerecetto H, Di Maio R, Ibarruri G, Seoane G, Denicola A, Peluffo G, et al. Synthesis and anti-trypanosomal activity of novel 5-nitro-2-furaldehyde and 5-nitrothiophene-2-carboxaldehyde semicarbazone derivatives. Il Farmaco. 1998;**53**:89-94. DOI: 10.1016/S0014-827X(97)00011-6

[75] Cerecetto H, Di Maio R, González M, Risso M, Sagrera G, Seoane G, et al. Synthesis and antitrypanosomal evaluation of E-isomers of 5-nitro-2-furaldehyde and 5-nitrothiophene-2-carboxaldehyde semicarbazone derivatives. Structure–activity relationships. European Journal of Medicinal Chemistry. 2000;**35**:343-350. DOI: 10.1016/S0223-5234(00)00131-8

[76] Williams DE, Yan JM. Point-charge models for molecules derived from least-squares fitting of the electric potential. Advances in Atomic and Molecular Physics. 1998;**23**:87-130. DOI: 10.1016/S00065-2199(08)60106-2

[77] Chirlian LE, Francl MM. Atomic charges derived from electrostatic potentials: A detailed study. Journal of Computational Chemistry. 1984;**8**:894-905. DOI: 10.1002/jcc.540080616

[78] Singh UC, Kollman PA. An approach to computing electrostatic charges for molecules. Journal of Computational Chemistry. 1984;**5**:129-145. DOI: 10.1002/jcc.5400502004

[79] Hyperchem 8.0.6, Inc. ChemPlus: Modular Extensions to HyperChem Release 8.06. Molecular Modeling for Windows 2008. Gainesville

[80] Tetko IV, Gasteiger J, Todeschini R, Mauri A, Livingstone D, Ertl P, et al. Virtual computational chemistry laboratory-design and description. Journal of Computer-Aided Molecular Design. 2005;**19**:453-463. DOI: 10.1007/s10822-005-8694-y

[81] Becke AD. Density-functional thermochemistry. III. The role of exact exchange. The Journal of Chemical Physics. 1993;**98**:5648-5652. DOI: 10.1063/1.464913

[82] Lee C, Yang W, Parr RG. Development of the colic-salvetti correlation-energy formula into a functional of the electron density. Physical Review B. 1988;**37**:785-789. DOI: 10.1103/PhysRevB.37.785

[83] Becke AD. Density-functional exchange-energy approximation with correct asymptotic behavior. Physical Review A. 1988;**38**:3098-3100. DOI: 10.1103/PhysRevA.38.3098

[84] Roothaan CC. New developments in molecular orbital theory. Reviews of Modern Physics. 1951;**23**:69-89. DOI: 10.1103/RevModPhys.23.69

[85] Dewar MJS, Zoebisch EG, Healy EF, Stewart JJP. Development and use of quantum mechanical molecular models. 76. AM1: A new general purpose quantum mechanical molecular model. Journal of the American Chemical Society. 1985;**107**:3902-3909. DOI: 390910.1021/ja00299a024

[86] Wu X, Thiel W, Pezeshki S, Lin H. Specific reaction path hamiltonian for proton transfer in water: Reparameterized semiempirical models. Journal of Chemical Theory and Computation. 2013;**9**:2672-2686. DOI: 10.1021/ct400224n

[87] Hehre WJ, Radom L, PvR S, Pople JA. Ab Initio Molecular Theory. New York: Wiley; 1986. DOI: 10.1002/jcc.540070314

[88] Ferreira MMC, Montanari CA, Gaudio AC. Seleção de variáveis em QSAR. Química Nova. 2002;**25**:439-448. DOI: 10.1590/S0100-40422002000300017

[89] Todeschini R, Consonni V. In: Mannhold R, Kubinyi H, Timmerman H, editors. Molecular Descriptors for Chemoinformatics. Vol I & II. Weinheim: Wiley-VCH; 2009. ISBN: 978-3-527-31852-0

[90] Karelson M, Victor S, Lobanov A, Katritzky R. Quantum-chemical descriptors in QSAR/QSPR studies. Chemical Reviews. 1996;**96**:1027-1043. DOI: 10.1021/cr950202r

[91] Gosav S, Praisler M, Dorohoi DO. ANN expert system screening for illicit amphetamines using molecular descriptors. Journal of Molecular Structure. 2007;**834**:188-194. DOI: 10.1016/j.molstruc.2006.12.059

[92] ScottiM,FernandesMA,FerreiraMJP, Esmereciano VP. Quantitative structure–activity relationship of sesquiterpene lactones with cytotoxic activity. Bioorganic & Medicinal Chemistry. 2007;**15**:2927-2934. DOI: 10.1016/j.bmc.2007.02.005

[93] Moriguchi I, Hirano S, Liu Q, Nakagome I, Matsushita Y. Simple method of calculating ocatanol/water partition coefficient. Chemical and Pharmaceutical Bulletin. 1992;**40**:127-130 ISSN: 1347-5223

[94] Paulino-Blumenfeld M, Hansz M, Hikici N. Electronic properties and free radical production by nitrofuran compounds. Free Radical Research Communications. 1992;**16**:207-215. DOI: 10.3109/10715769209049174

Chapter 5

Chemical Reactivity Properties and Bioactivity Scores of the Angiotensin II Vasoconstrictor Octapeptide

Norma Flores-Holguín, Juan Frau and Daniel Glossman-Mitnik

Abstract

Eight density functionals, CAM-B3LYP, LC-ωPBE, M11, MN12SX, N12SX, ωB97, ωB97X, and ωB97XD, in connection with the Def2TZVP basis set were assessed together with the SMD solvation model for the calculation of the molecular and chemical reactivity properties of the angiotensin II vasoconstrictor octapeptide in the presence of water. All the chemical reactivity descriptors for the systems were calculated via conceptual density functional theory (CDFT). The potential bioavailability and druggability as well as the bioactivity scores for angiotensin II were predicted through different methodologies already reported in the literature which have been previously validated during the study of different peptidic systems.

Keywords: angiotensin II, conceptual DFT, chemical reactivity, drug-likeness features, bioactivity scores

1. Introduction

In order to consider peptides and related compounds as the starting point for the development of medical drugs, it is mandatory to acquire a knowledge about their chemical reactivity properties as well as the bioactivity associated with them. From the basics of medicinal chemistry, it is known that drugs exert their effect by interacting with the active site of a receptor which is generally a protein [1]. These interactions rely on the different kinds of bindings between the pharmacophore and the chemical groups present in the active site and thus intimately related to their chemical reactivity from a molecular perspective [2, 3]. One of the most powerful tools to understand the chemical reactivity of interacting molecular systems within computational chemistry is probably the conceptual density functional theory (CDFT) [4, 5], also called chemical reactivity theory, which allows to accomplish this task by resorting to several global and local descriptors which are in turn related to variations in the electronic densities of the studied systems.

On the basis of the previous considerations, the objective of this work is to study the chemical reactivity of an octapeptide known as angiotensin II that acts constricting the blood vessels and retaining the fluid in the kidneys [1], using the techniques of the conceptual DFT, determining their global reactivity properties, that is, of the molecule as a whole. Moreover, during the process of the development of new drugs, there is a need to learn about the drug-like properties of the involved molecular systems [6]. Thus, the descriptors of bioavailability and bioactivity (bioactivity scores) will be calculated through different procedures described in the literature [7, 8] trying to relate them with the calculated conceptual DFT descriptors.

2. Computational methodology

In the same way as we have proceeded in our recent studies [9–16], the computational tasks in this work have been done by considering the popular Gaussian 09 software [17]. Following the conclusions obtained from those studies, eight density functionals have been chosen, CAM-B3LYP, LC-ωPBE, M11, MN12SX [18], N12SX, ωB97, ωB97X, and ωB97XD, because they can be considered to be well-behaved for our purposes according to our proposed KID (for Koopmans in DFT) criteria [19–23] related to the approximate validity of the Koopmans' theorem within DFT [19–23]. For the calculation of the electronic properties, several model chemistries have been considered, based on the mentioned density functionals in connection with the Def2TZVP basis set, while a smaller Def2SVP was considered for the prediction of the most stable structures [24, 25]. In order to obtain accurate results, all calculations were performed using water, which is the universal biological solvent, simulated with the SMD model [26].

3. Results and discussion

The molecular structures of the conformers of the angiotensin II vasoconstrictor octapeptide graphically presented in **Figure 1** were optimized in the gas phase by means of the DFTBA model available in the software and then reoptimized with the eight density functionals described previously, the Def2SVP basis set, and water as the solvent. The calculation of the electronic properties was performed by using the same model chemistries but changing the basis set with the Def2TZVP one.

In order to verify the fulfillment of our proposed KID procedure, it is necessary to perform a comparison of the orbital energies with the results obtained by means of the vertical I and A through the ΔSCF criterium. To this end, the three main descriptors are linked by ε_H with $-I$, ε_L with $-A$, and their behavior in describing the HOMO-LUMO gap as $J_I = |\varepsilon_H + E_{gs}(N-1) - E_{gs}(N)|$,

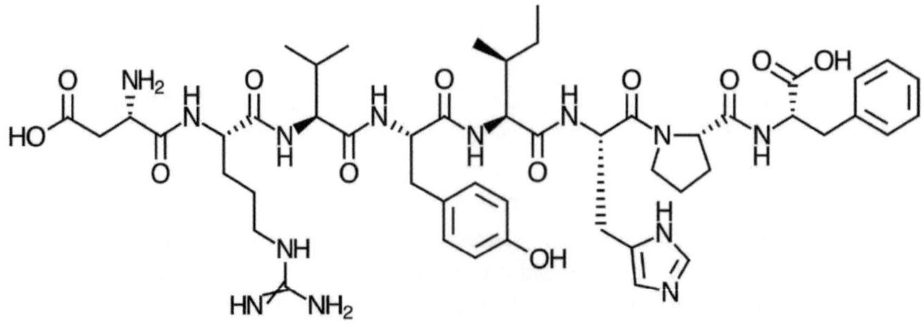

Figure 1.
Graphical sketch of the angiotensin II molecule.

	Eo	E+	E−	HOMO	LUMO
CAM-B3LYP	−1887.465	−1887.246	−1887.489	−7.462	0.828
LC-wBPE	−1887.192	−1886.966	−1887.223	−8.786	1.767
M11	−1887.317	−1887.090	−1887.345	−8.601	1.582
MN12SX	−1886.668	−1886.440	−1886.699	−6.164	−0.832
N12SX	−1887.505	−1887.288	−1887.531	−5.881	−0.679
ωB97	−1888.093	−1887.871	−1888.118	−8.658	1.890
ωB97X	−1887.933	−1887.711	−1887.959	−8.474	1.724
ωB97XD	−1887.814	−1887.592	−1887.840	−8.087	1.374
	SOMO	**J_I**	**J_A**	**J_{HL}**	**ΔSL**
CAM-B3LYP	−2.205	1.497	1.498	2.117	3.033
LC-wBPE	−3.509	2.635	2.619	3.715	5.276
M11	−3.124	2.412	2.333	3.356	4.706
MN12SX	−0.869	0.021	0.017	0.028	0.038
N12SX	−0.785	0.000	0.053	0.053	0.106
ωB97	−3.303	2.619	2.575	3.673	5.192
ωB97X	−3.144	2.432	2.410	3.424	4.868
ωB97XD	−2.809	2.059	2.073	2.922	4.183

Table 1.
Total electronic energies of angiotensin II (in au) for the neutral and charged species, the corresponding orbital energies (in eV), and the KID-related descriptors obtained with the five density functionals, the Def2TZVP basis set, and water as the solvent.

$J_A = |\varepsilon_L + E_{gs}(N) - E_{gs}(N+1)|$, and $J_{HL} = \sqrt{J_I^2 + J_A^2}$. Another descriptor, ΔSL (the difference between the SOMO and the LUMO), was also designed to guide in verifying the accuracy of the approximation [9–15]. The results of this analysis are presented in **Table 1**.

The overall conclusion that can be extracted from the inspection of the results presented in **Table 1** is that, in agreement with our previous studies on melanoidins and peptides, the model chemistries involving the MN12SX and N12SX density functionals are the best for verifying our proposed criteria of well-behavior.

3.1 Calculation of the global reactivity descriptors

By taking into account the KID procedure presented in our previous works together with the finite difference approximation, the global reactivity descriptors can be expressed as

Electronegativity	$\chi = -\frac{1}{2}(I + A) \approx \frac{1}{2}(\varepsilon_L + \varepsilon_H)$	[4, 5]
Global hardness	$\eta = (I - A) \approx (\varepsilon_L - \varepsilon_H)$	[4, 5]
Electrophilicity	$\omega = \frac{\mu^2}{2\eta} = \frac{(I+A)^2}{4(I-A)} \approx \frac{(\varepsilon_L+\varepsilon_H)^2}{4(\varepsilon_L-\varepsilon_H)}$	[27]
Electrodonating power	$\omega^- = \frac{(3I+A)^2}{16(I-A)} \approx \frac{(3\varepsilon_H+\varepsilon_L)^2}{16\eta}$	[28]
Electroaccepting power	$\omega^+ = \frac{(I+3A)^2}{16(I-A)} \approx \frac{(\varepsilon_H+3\varepsilon_L)^2}{16\eta}$	[28]
Net electrophilicity	$\Delta\omega^{\pm} = \omega^+ - (-\omega^-) = \omega^+ + \omega^-$	[29]

where I is the ionization potential and A the electronic affinity, while ε_H and ε_L are the energies of the HOMO and LUMO, respectively.

The results for the global reactivity descriptors for the angiotensin II octapeptide based on the values of the HOMO and LUMO energies calculated with the MN12SX and N12SX density functionals are presented in **Table 2**.

As expected from the molecular structure of this peptide, its electrodonating ability is more important that its electroaccepting character. It can be seen that MN12SX and N12SX density functionals (which verify the KID criteria) give results different than those obtained from the calculation with the other three density functionals.

3.2 Bioactivity scores

The molecular properties that are related to the concept of drug-likeness and in particular those associated with the criteria proposed by Lipinski et al. [30, 31] for the prediction of oral bioavailability have been calculated by feeding the corresponding SMILES notations into the Molinspiration software readily available online (Slovensky Grob, Slovak Republic: https://www.mol inspiration.com). The results are presented in **Table 3**.

However, what the Lipinski's rule of five really measures is the oral bioavailability of a potential drug because this is the desired property for a molecule having drug-like character. Then, a different approach was followed by considering similarity searches in the chemical space of compounds with structures that can be

	Electronegativity (χ)	**Chemical hardness (η)**	**Electrophilicity (ω)**
MN12SX	3.3286	4.9685	1.1150
N12SX	3.1472	4.7664	1.0391
	Electrodonating power (ω^-)	**Electroaccepting power (ω^+)**	**Net electrophilicity ($\Delta\omega^{\pm}$)**
MN12SX	2.4725	1.1286	3.6011
N12SX	2.3225	1.0468	3.3693

Table 2.
Global reactivity descriptors for the angiotensin II molecule calculated with the MN12SX and N12SX density functionals with the Def2TZVP basis set and the SMD solvation model using water as the solvent.

Molecule	**Angiotensin II**
miLogP	−3.91
TPSA	406.33
nAtoms	75
nON	25
nOHNH	16
nviol	3
nrotb	30
volume	955.57
MW	1046.20

Table 3.
Molecular properties of the angiotensin II peptide calculated to verify the Lipinski's rule of five.

Molecule	Angiotensin II
GPCR ligand	−3.59
Ion channel modulator	−3.74
Kinase inhibitor	−3.78
Nuclear receptor ligand	−3.85
Protease inhibitor	−3.25
Enzyme inhibitor	−3.67

Table 4.
Bioactivity scores of the angiotensin II molecule calculated on the basis of GPCR ligand, ion channel modulator, nuclear receptor ligand, kinase inhibitor, protease inhibitor, and enzyme inhibitor interactions.

compared to those that are being studied and with known pharmacological properties. The same software was used for the calculation of the bioactivity scores which are a measure of the ability of the potential drug to interact with the different receptors, that is, to act as GPCR ligands or kinase inhibitors, to perform as ion channel modulators, or to interact with enzymes and nuclear receptors. The values of the bioactivity scores for angiotensin II are presented in **Table 4**.

These bioactivity scores for organic molecules can be interpreted as active (when the bioactivity score > 0), moderately active (when the bioactivity score lies between −5.0 and 0.0), and inactive (when the bioactivity score < −5.0). The angiotensin II peptide was found to be moderately bioactive toward the protease inhibitor and the GPCR ligand considered in the study.

4. Conclusions

In this chapter we have presented a new study performed on the chemical reactivity of the angiotensin II vasoconstrictor octapeptide based on the conceptual DFT as a tool to explain the molecular interactions.

The knowledge of the values of the global descriptors of the molecular reactivity of angiotensin II could be useful in the development of new drugs based on this compound or some analogs.

Finally, the molecular properties related to bioavailability and drug-likeness have been predicted using a proven methodology already described in the literature, and the descriptors used for the quantification of the bioactivity allowed to characterize the studied molecule as being moderately bioactive toward the protease inhibitor and the GPCR ligand considered in this study.

Acknowledgements

Norma Flores-Holguín and Daniel Glossman-Mitnik are researchers of CIMAV and CONACYT from which partial support is gratefully acknowledged. Daniel Glossman-Mitnik conducted this work while being a visiting lecturer at the University of the Balearic Islands. This work was also cofunded by the Ministerio de Economía y Competitividad (MINECO) and the European Fund for Regional Development (FEDER).

Conflict of interest

The authors declare no conflict of interest regarding the publication of this chapter.

Author details

Norma Flores-Holguín[1†], Juan Frau[2†] and Daniel Glossman-Mitnik[1*†]

1 Centro de Investigación en Materiales Avanzados, Departamento de Medio Ambiente y Energía, Laboratorio Virtual NANOCOSMOS, Chihuahua, Mexico

2 Departament de Química, Universitat de les Illes Balears, Palma de Mallorca, Spain

*Address all correspondence to: daniel.glossman@cimav.edu.mx

†These authors contributed equally.

References

[1] Patrick GL. An Introduction to Medicinal Chemistry. Oxford, UK: Oxford University Press; 2013

[2] Rekka EA, Kourounakis PN. Chemistry and Molecular Aspects of Drug Design and Action. Boca Raton: CRC Press; 2008

[3] N'aray-Szabó G'a, Warshel A. Computational Approaches to Biochemical Reactivity. New York: Kluwer Academic Publishers; 2002

[4] Parr RG, Yang W. Density-Functional Theory of Atoms and Molecules. New York: Oxford University Press; 1989

[5] Geerlings P, De Proft F, Langenaeker W. Conceptual density functional theory. Chemical Reviews. 2003;**103**:1793-1873

[6] Stromgaard K, Krogsgaard-Larsen P, Madsen U. Textbook of Drug Design and Discovery. Boca Raton, FL: CRC Press/Taylor and Francis Group; 2017

[7] Gupta GK, Kumar V. Chemical Drug Design. Berlin: Walter de Gruyter GmbH; 2016

[8] Gore M, Jagtap UB. Computational Drug Discovery and Design. New York: Springer Science+Business Media, LLC; 2018

[9] Frau J, Glossman-Mitnik D. Molecular reactivity and absorption properties of melanoidin blue-G1 through conceptual DFT. Molecules. 2018;**23**(3):559-515

[10] Frau J, Glossman-Mitnik D. Conceptual DFT study of the local chemical reactivity of the dilysyldipyrrolones A and B intermediate melanoidins. Theoretical Chemistry Accounts. 2018; **137**(5):1210

[11] Frau J, Glossman-Mitnik D. Conceptual DFT study of the local chemical reactivity of the colored BISARG melanoidin and its protonated derivative. Frontiers in Chemistry. 2018;**6**(136):1-9

[12] Frau J, Glossman-Mitnik D. Molecular reactivity of some Maillard reaction products studied through conceptual DFT. Contemporary Chemistry. 2018;**1**(1):1-14

[13] Frau J, Glossman-Mitnik D. Computational study of the chemical reactivity of the blue-M1 intermediate melanoidin. Computational and Theoretical Chemistry. 2018;**1134**:22-29

[14] Frau J, Glossman-Mitnik D. Chemical reactivity theory applied to the calculation of the local reactivity descriptors of a colored Maillard reaction product. Chemical Science International Journal. 2018;**22**(4):1-14

[15] Frau J, Glossman-Mitnik D. Blue M2: An intermediate melanoidin studied via conceptual DFT. Journal of Molecular Modeling. 2018;**24**(138):1-13

[16] Frau J, Flores-Holguín N, Glossman-Mitnik D. Chemical reactivity properties, pKa values, AGEs inhibitor abilities and bioactivity scores of the mirabamides A–H peptides of marine origin studied by means of conceptual DFT. Marine Drugs. 2018; **16**(9):302-319

[17] Frisch MJ, Trucks GW, Schlegel HB, Scuseria GE, Robb MA, Cheeseman JR, et al. Gaussian 09 Revision E.01. Wallingford, CT: Gaussian Inc.; 2016

[18] Peverati R, Truhlar DG. Screened-exchange density functionals with broad accuracy for chemistry and solid-state physics. Physical Chemistry Chemical Physics. 2012;**14**(47):16187-16191

[19] Borghi G, Ferretti A, Nguyen NL, Dabo I, Marzari N. Koopmans-compliant functionals and their performance against reference molecular data. Physical Review B. 2014;**90**(7):1

[20] Dabo I, Ferretti A, Poilvert N, Li Y, Marzari N, Cococcioni M. Koopmans' condition for density-functional theory. Physical Review B. 2010;**82**(11):115121

[21] Kar R, Song J-W, Hirao K. Long-range corrected functionals satisfy Koopmans' theorem: Calculation of correlation and relaxation energies. Journal of Computational Chemistry. 2013;**34**(11):958-964

[22] Salzner U, Baer R. Koopmans' springs to life. The Journal of Chemical Physics. 2009;**131**(23):231101

[23] Vanfleteren D, Van Neck D, Ayers PW, Morrison RC, Bultinck P. Exact ionization potentials from wavefunction asymptotics: The extended Koopmans' theorem, revisited. The Journal of Chemical Physics. 2009;**130**(19):194104

[24] Weigend F, Ahlrichs R. Balanced basis sets of split valence, triple zeta valence and quadruple zeta valence quality for H to Rn: Design and assessment of accuracy. Physical Chemistry Chemical Physics. 2005;**7**: 3297-3305

[25] Weigend F. Accurate Coulomb-fitting basis sets for H to R. Physical Chemistry Chemical Physics. 2006;**8**: 1057-1065

[26] Marenich AV, Cramer CJ, Truhlar DG. Universal solvation model based on solute electron density and a continuum model of the solvent defined by the bulk dielectric constant and atomic surface tensions. Journal of Physical Chemistry B. 2009;**113**:6378-6396

[27] Parr RG, Szentpaly LV, Liu SB. Electrophilicity index. Journal of the American Chemical Society. 1999;**121**: 1922-1924

[28] Gázquez JL, Cedillo A, Vela A. Electrodonating and electroaccepting powers. Journal of Physical Chemistry A. 2007;**111**(10):1966-1970

[29] Chattaraj PK, Chakraborty A, Giri S. Net electrophilicity. Journal of Physical Chemistry A. 2009;**113**(37): 10068-10074

[30] Lipinski CA, Lombardo F, Dominy BW, Feeney PJ. Experimental and computational approaches to estimate solubility and permeability in drug discovery and development settings. Advanced Drug Delivery Reviews. 2001;**46**:3-26

[31] Leeson P. Drug discovery: Chemical beauty contest. Nature. 2012;**481**(7382): 455-456

Chapter 6

Chemoinformatic Approach: The Case of Natural Products of Panama

Dionisio A. Olmedo and José L. Medina-Franco

Abstract

Chemoinformatic analysis was used to characterize a compound database of natural products from Panama and other reference collections. Data mining allowed to compare drug-likeness properties with public and commercial software and to achieve a statistical analysis of the physicochemical properties. Visualization of the chemical space in 3D indicates a high structural similarity. Molecular flexibility and complexity were evaluated using 2D descriptors, whereas the molecular scaffold was obtained using the Murcko method, and these showed few differences between the explored data set. In this chapter, we also present and discuss an example of the application of the chemoinformatic approach using the concept of modeling the activity landscape to study the structure-activity relationships (SARs) of compounds with activity against *Plasmodium falciparum*.

Keywords: chemoinformatic, complexity, data mining, physicochemical properties, scaffold

1. Introduction

Natural products (NPs) and their derivatives constitute a significant fraction of approved drugs [1–3], bioactive compounds [4–8], and lead compounds for drug discovery [9]. NP fragment has been used to guide the synthesis of bioactive compounds and generate BIOS combinatorial libraries [10–15]. NPs have structures with different substituent patterns, giving rise to different biological activities for compounds with very similar structures [16–19]. These bioactive metabolites have greater affinity for biological targets and, overall, may have better bioavailability than synthetic compounds, and the presence of pan-assay interference compounds (PAIN) is less frequent in this type of product [20]. The chemoinformatic analysis of several databases of NPs developed by academic institutions and private companies [21] has been carried out in different countries. Thus, the following databases were obtained: BIOFACQUIM [22], CIFPMA [23], NuBBE [24, 25], NANPDB [26], TCM [27], HIT [28], and NPACT [29]. The application of chemoinformatic tools involves the generation, manipulation, and analysis of data set of chemical substances. This allows us through mathematical calculations to order, develop, and evaluate structural information that can be visualized in 2D and 3D [30]. The determination of the physicochemical properties carried out on different databases of NPs and principal component analysis (PCA) was used as an approximation to display the chemical spaces [22–24, 31–37].

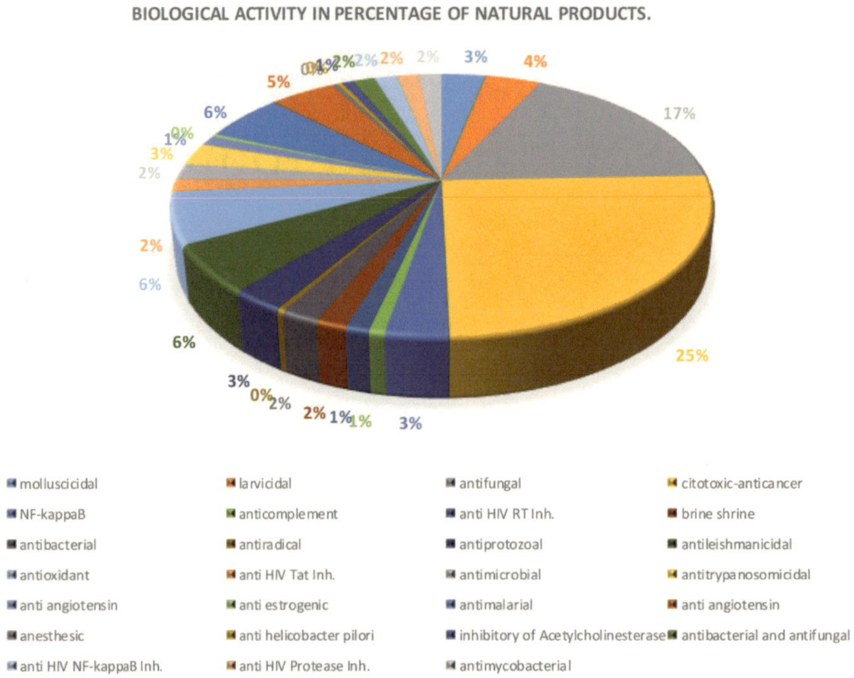

Figure 1.
Biological endpoints and targets in which natural products from Panama present bioactivity.

Computational exploration of NPs has increased in recent years, giving greater relevance to studies that include structural diversity metrics calculated with parameters based on distances such as Euclidean distance, Manhattan distances, and Cosine distance. Other criteria are based on circular fingerprint (ECFP-4, ECFP-6) [22–24, 38–45] and fingerprint based on substructure (MACCS, PubChem) [22–24, 39–45]. Another metric used in NPs is the comparison by similarity that uses the Tanimoto index/Tanimoto coefficient [22–24, 45–49].

In this study, the molecular scaffolds of natural products have been obtained using the Murcko method [22–24, 50–57]. Meanwhile, the molecular complexity is frequently evaluated by descriptors in 2D such as fraction of sp^3 hybridized carbons (Fsp^3) [23], fraction of chiral centers (FCC) [23], and globularity [22–24, 58–63].

An update of the Natural Products Database from the University of Panama (UPMA) containing 454 compounds (Unpublished data) has been evaluated against different therapeutic targets such as cytotoxicity bioassay in cell lines, antifungal assay in vitro, parasites of tropical diseases (*Leishmania* sp., *Plasmodium falciparum*, and *Trypanosoma cruzi*), and the bioassay against HIV-1 virus, demonstrating an inhibitor effect on protease, reverse transcriptase, nuclear factor NFkappaB, and Tat protein affecting the viral replication. These are the most significant biological targets in which the natural products from Panama present bioactivity. The values of their biological activities are represented as percentages in **Figure 1**.

2. Application of chemoinformatic antimalarial databases: case of natural products from Panama

2.1 Preparation curated and processing of data set

In this chapter, we present a chemoinformatic analysis of natural products with antimalarial activities (in vitro), expressed as pIC_{50} against sensitive and resistant strains. Databases of natural products with antimalarial activity (NPAs) were constructed in-house by reviewing published articles including those compounds that were isolated and characterized by spectroscopic techniques of nuclear magnetic resonance. Around 1312 compounds were compared to 8 reference data sets: an open database, DrugBank (antimalarial drug), European Bioinformatics Institute. (CHEMBL drug indications) (antimalarial activities), Open Source Drug Discovery (OSDD) Malaria, Malaria Box (Medicines for Malaria Venture (MMV)), St. Jude Children's Research Hospital (St. Jude), Novartis (GNF Malaria Box), and GlaxoSmithKline (GSK) Tres Cantos antimalarial set. All data sets were curated using the "Wash" function implemented in the Molecular Operating Environment (MOE2018.0101) software [64]. The structure of the studied compounds was represented by simplified molecular input line entry system (SMILES) notation, thus obtaining 20,364 unique molecules that are summarized in **Table 1**. The difference between initial compounds and unique compounds is due to the fact that during the data preparation (curation process), the duplicate compounds are eliminated, those that have positive or negative partial loads have neutralized their protonation states, the metals are disconnected, and the energy is minimized using the molecular mechanistic force field (MMFF94). The result of the data curation is the reduction of the initial number of molecules present in the databases evaluated in this work.

2.2 Molecular descriptors

The descriptors of physicochemical properties, hydrogen bond acceptors (HBAs), hydrogen bond donors (HBDs), number of rotatable bonds (NRBs), the octanol/water partition coefficient (logP), topological polar surface area (TPSA),

Databases	Initial compounds	Unique compounds	Source
Natural Products Antimalarial (NPAs)	1353	1312	Databases of NP in house
DrugBank Version 5.0. (Drug Antimalarial)	26	4	https://www.drugbank.ca
European Bioinformatics Institute. (CHEMBL Drugs Indications) (Antimalarial activities	27	24	[https://www.ebi.ac.uk/chembl]
Open Source Drug Discovery (OSDD) Malaria	93	88	http://opensourcemalaria.org/
Malaria Box-Medicine of Malaria Venture (MMV)	124	124	https://www.ebi.ac.uk/chembl/malaria/source
St. Jude Children's Research Hospital's	1.478	1.478	https://www.ebi.ac.uk/chemblntd
Novartis-GNF Malaria Box	4.878	4.868	Available in: https://www.ncbi.nlm.nih.gov/pmc/articles/PMC3941073/ Available in: https://www.ebi.ac.uk/chemblntd
GlaxoSmithKline Tres Cantos Antimalarial	12.470	12.466	Open Source Malaria (GSK-TCMDC). Available in: https://www.ebi.ac.uk/chemblntd

Table 1.
Databases analyzed with chemoinformatic tools.

and molecular weight (MW), or others such as molar refractivity, are important physicochemical parameters for quantitative structure-activity relationship (QSAR) analysis. These molecular descriptors are based on Lipinski's rule and Verger's rule regarding the prediction of the pharmacological similarity of orally active pharmacological potential [65–67]. The statistical analysis of the physicochemical properties was realized with RStudio Software 1.0.136 AGPL [68].

2.3 3D visualization of chemical space of compounds with antimalarial activity

PCAs were done with MOE software [64], and the dominant characteristics are expressed as covariance and visualized with the corresponding 2D or 3D graphic score plot with DataWarrior program v. 5.0 [69]. **Figures 2–8** showed the distribution of different compounds with antimalarial activities in the chemical spaces.

In **Figures 2–8** we observed that NPs, drugs, and synthetic compounds occupy, in general, similar chemical space and are overlapping in most of the evaluated databases.

2.4 Molecular diversity based on fingerprints

Three binary molecular fingerprints were calculated with RStudio package rcdk: Extended connectivity fingerprints with diameter 4 (ECFP-4) for similarity searching, molecular access system (MACCS) keys of 166 bits for determining similarity and molecular diversity, and PubChem keys of 881 bits for encoding molecular fragment information [42–44]. The similarity of fingerprints by structural pairs of compounds was calculated with the Tanimoto coefficient and analyzed with the cumulative distribution function (CDF). This approach has been used to calculate, measure, and represent the molecular variety of compound data sets [23].

Figures 9–11 show the CDFs of the pairwise similarity of the different data sets evaluated with Tanimoto coefficient and ECPF-4, MACCS keys, and PubChem fingerprints, respectively.

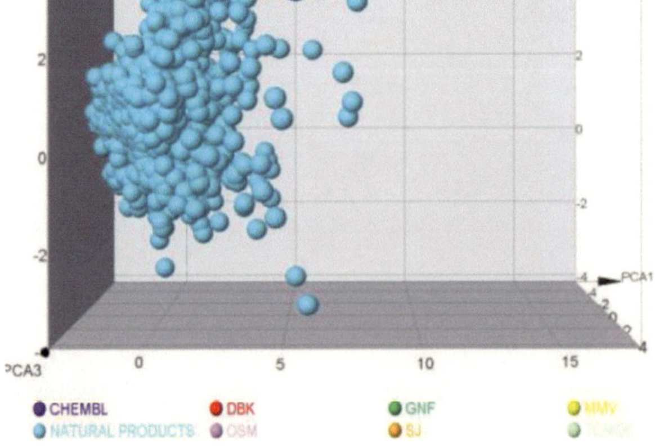

Figure 2.
3D visualization of the chemical space of natural product databases.

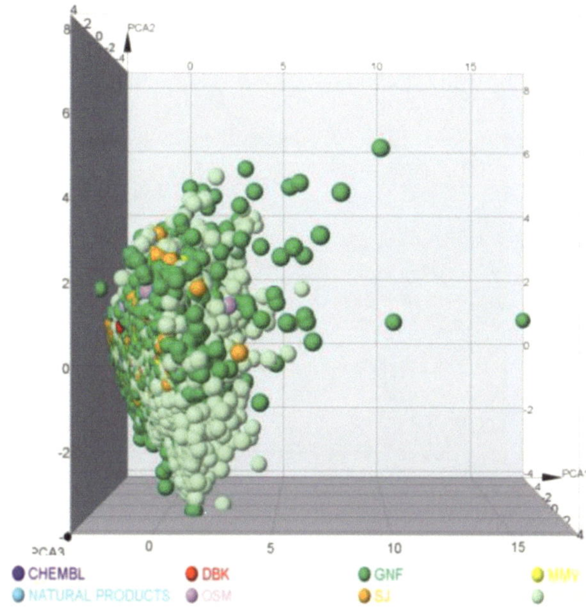

Figure 3.
3D visualization of the chemical space of synthetic compounds.

Figures 9–11 provide information on the structural diversity of the six databases. Similar approach has been previously published [23]; the curves obtained with ECFP-4 did not prove to be a suitable fingerprint representation for these data sets. In the three similarity graphs based on fingerprints, it is shown that the database of natural products with antimalarial activity, OMS, and MMV has the lowest molecular diversity, while GSK DB was the most diverse.

In **Tables 2–4**, the statistical values of the pairwise Tanimoto similarity with the data sets analyzed are shown. In these tables, CHEMBL and DrugBank databases are excluded from our analysis, due to the small amount of data.

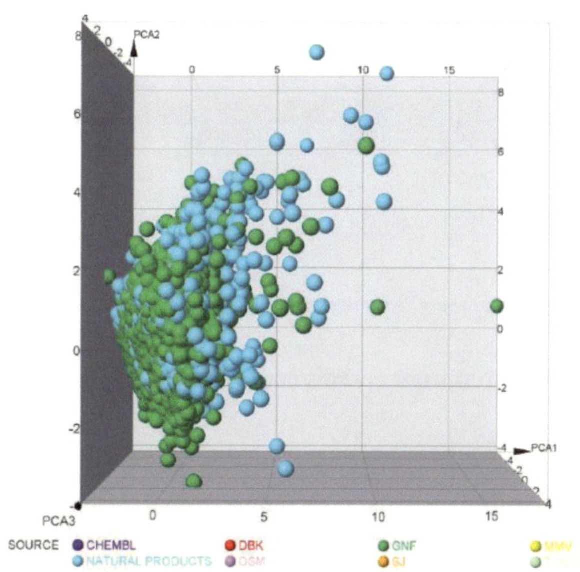

Figure 4.
3D visualization of the chemical spaces of natural products and GNF DBs.

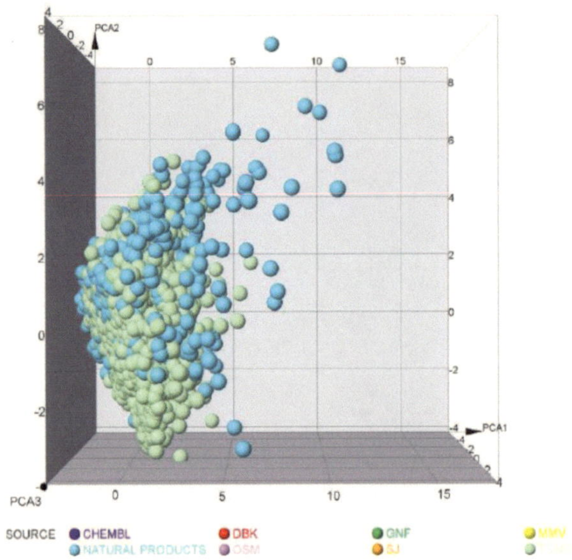

Figure 5.
3D visualization of the chemical spaces of natural products and TCMDC DBs.

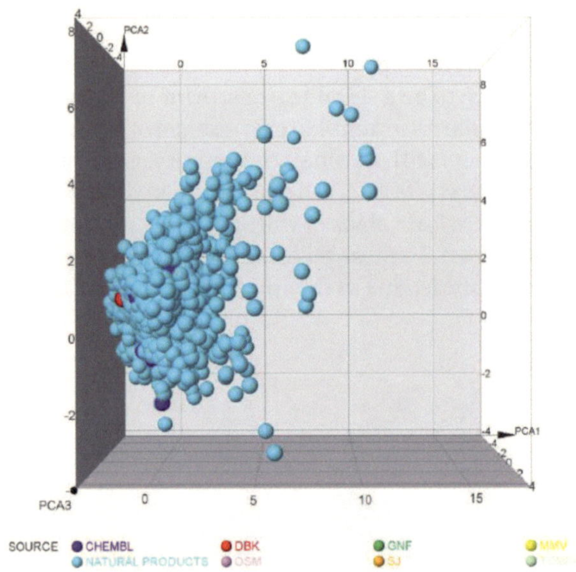

Figure 6.
3D visualization of the chemical spaces of natural products and DBK DBs.

2.5 Molecular scaffolds: content and diversity

2.5.1 Scaffold content

Murcko scaffolds were calculated with the program Molecular Equivalent Indices (MEQI) [50, 51] and DataWarrior program [69]. MEQI has been used to obtain the codes corresponding to the chemotypes most frequently analyzed in the databases. [23, 45, 52–55]. The distribution and diversity of the molecular scaffolds present in the data sets were calculated and analyzed using the cyclic system

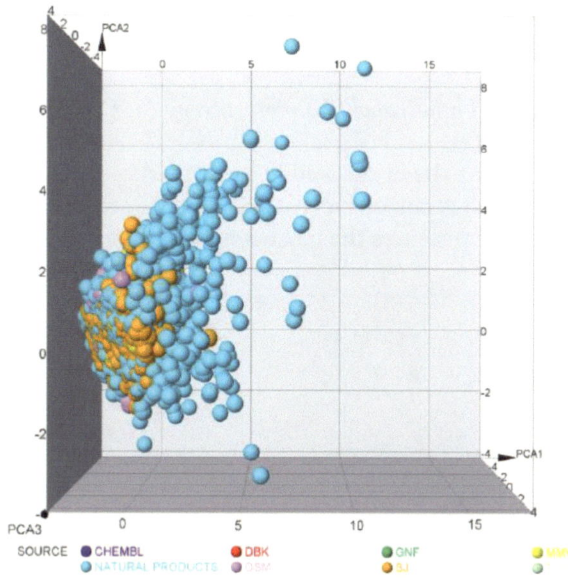

Figure 7.
3D visualization of the chemical spaces of natural products, OSM and St. Jude.

retrieval (CSR) curves [42]. These curves were obtained by plotting the fraction of scaffold and the fraction of compounds that contain cyclic systems [43, 44].

Table 5 indicates that the MMV DB (0.491) was the most diverse in scaffold content taken as reference the F_{50} values compared to the data set from GSK (0.183), NPs (0.168), and GNF (0.161), respectively. CSR curves on **Figure 12** further confirm the relative scaffold variety of the eight databases. The analysis of area under curve (AUC) metrics associated with the CSR curves is reported in **Table 5**. The CSR curves showed that MMV has more variety in scaffold content with AUC value of 0.507. In contrast OSM, NPs, GNF, GSK, St. Jude, and CHEMBL were the least diverse (e.g., AUC scores of 0.745, 0.712, 0.705, 0.698, 0.655 and 0.607,

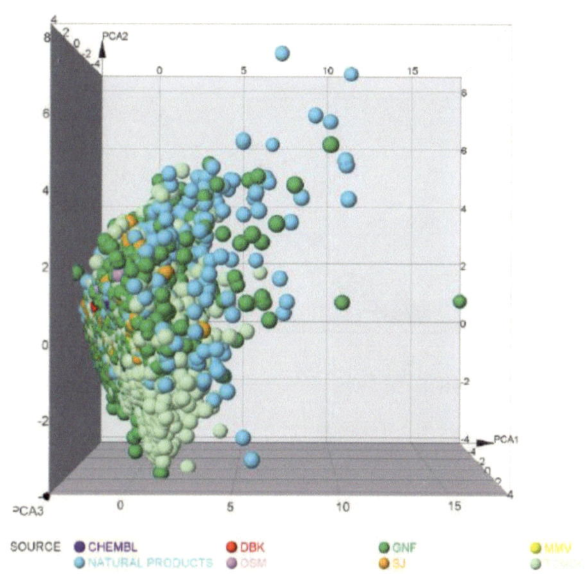

Figure 8.
3D visualization of the chemical spaces of all databases.

respectively). The CSR curves provide information on the diversity of the most frequent scaffolds in all databases.

2.5.2 Shannon entropy (SE) and scaled Shannon entropy (SSE)

The Shannon entropy has been adapted to measure the scaffold diversity based on the (**N**) number of most recurrent scaffolds [70]. The scaled Shannon entropy is a normalized value that measures the most common chemotypes present in a

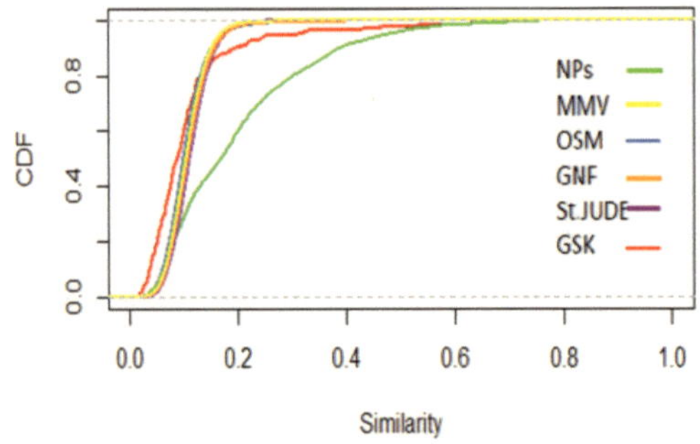

Figure 9.
Curve for cumulative frequency distribution (CFD) based on ECFP-4.

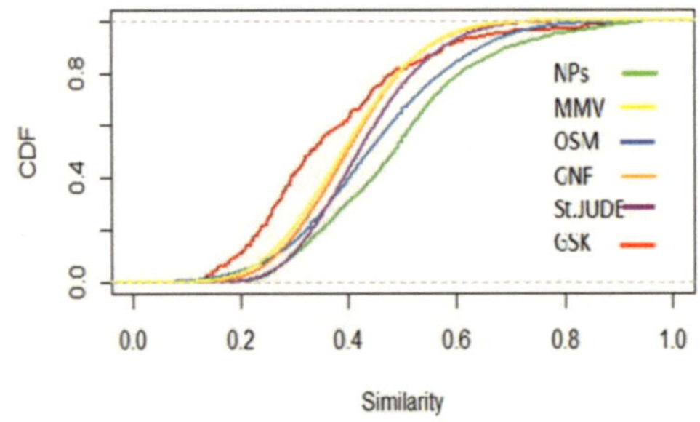

Figure 10.
Curve for cumulative frequency distribution based on MACCS keys.

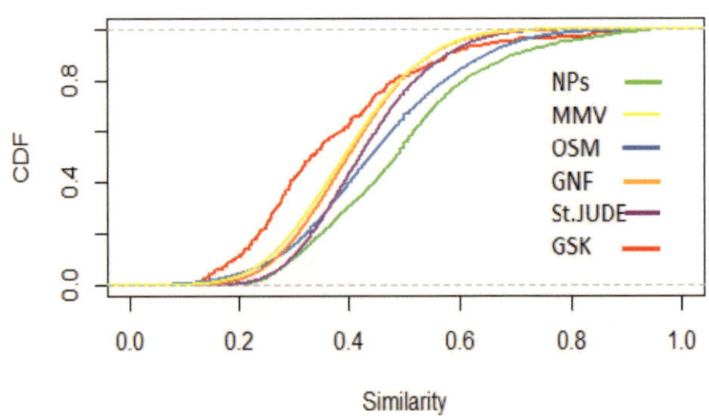

Figure 11.
Curve for cumulative frequency distribution based on PubChem.

database. Thus, SSE closer to 1 indicates higher scaffold diversity, while SSE closer to zero (0) indicates lower diversity. In this study, we calculated the SSE for values ranging from **N** = 10 to **N** = 40.

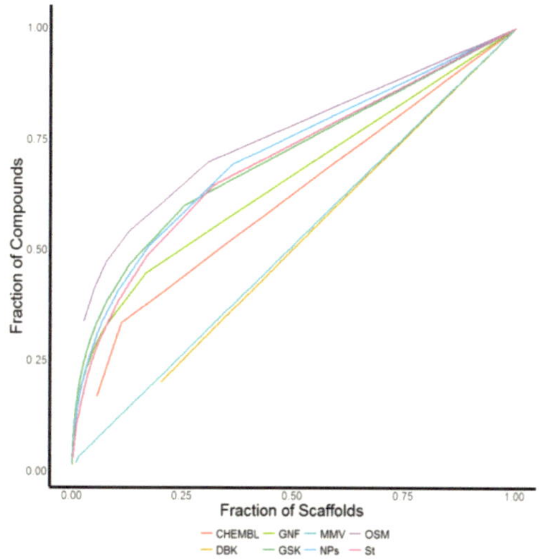

Figure 12.
Cyclic system retrieval curves for all databases evaluated in this study.

Similarity ECFP-4/Tanimoto coefficient						
DBs	**Min.**	**1st Qu.**	**Median**	**Mean**	**3rd Qu.**	**Max.**
GSK	0.01724	0.05789	0.08844	0.11490	0.12245	0.82353
NPs	0.00000	0.07826	0.09910	0.10565	0.12389	1.00000
OSM	0.00000	0.07826	0.09917	0.10607	0.12397	1.00000
MMV	0.00000	0.07826	0.09924	0.10615	0.12403	1.00000
ST JUDE	0.00000	0.08197	0.10345	0.10980	0.12857	1.00000
GNF	0.00000	0.08209	0.10345	0.10772	0.12739	1.00000

Table 2.
The statistical values of the similarity of the Tanimoto coefficient with ECFP-4.

Similarity MACCS keys/Tanimoto coefficient						
DBs	**Min.**	**1st Qu.**	**Median**	**Mean**	**3rd Qu.**	**Max.**
GSK	0.07813	0.25682	0.33333	0.37009	0.45581	0.92683
NPs	0.00000	0.34426	0.43636	0.44673	0.54545	1.00000
OSM	0.00000	0.34483	0.43636	0.44693	0.54545	1.00000
MMV	0.00000	0.34483	0.43636	0.44677	0.54412	1.00000
ST JUDE	0.00000	0.33333	0.41250	0.42313	0.50000	1.00000
GNF	0.00000	0.31746	0.39437	0.39999	0.47619	1.00000

Table 3.
The statistical values of the similarity of the Tanimoto coefficient with MACCS keys.

Figure 13 shows a histogram with the distribution of the 40 most populated scaffolds in NPAs. The histogram includes the corresponding chemotype code. The comparison of the scaffolds of the NPAs allowed the identification of the 68MBD chemotype as one of the most active compounds in this database.

Similarity PubChem/Tanimoto coefficient						
DBs	**Min.**	**1st Qu.**	**Median**	**Mean**	**3rd Qu.**	**Max.**
GSK	0.08125	0.24500	0.37555	0.40263	0.54002	1.00000
NPs	0.03684	0.32298	0.43802	0.46184	0.58621	1.00000
OSM	0.03684	0.32340	0.43902	0.46253	0.58730	1.00000
MMV	0.03684	0.32444	0.44033	0.46321	0.58791	1.00000
ST JUDE	0.03684	0.38224	0.47143	0.47624	0.56195	1.00000
GNF	0.00000	0.40598	0.48117	0.47800	0.55446	1.00000

Table 4.
The statistical values of the similarity of the Tanimoto coefficient with PubChem.

DBs	Number of Compounds (M)	Unique chemotypes (N)	FN/M	NSING	FNSING/M	FNSING/NS	AUC	F_{50}
NPs	1298	629	0.4846	400	0.3082	0.6359	0.7125	0.1685
DBK	5	5	1.0000	5	1.0000	1.0000	0.4800	0.4000
CHEMBL	24	18	0.7500	16	0.6667	0.8889	0.6072	0.3333
OSM	89	39	0.4382	27	0.3034	0.6923	0.7453	0.1025
MMV	124	122	0.9839	120	0.9677	0.9836	0.5079	0.4918
St. JUDE	915	479	0.5235	325	0.3552	0.6785	0.6551	0.2474
GNF	4860	3229	0.6644	2690	0.5535	0.8331	0.7054	0.1615
GSK	12,463	6703	0.5378	5009	0.4019	0.7473	0.6982	0.1837

M = number of molecules in the BD, N = number of chemotypes or substructures, FN/M = chemotype diversity fraction, NSING = singleton number, FNSING/M = singleton fraction between total molecules, FNSING/N = fraction of singleton among total chemotypes, AUC = area under the curve, F50 = fraction of chemotype required to recover 50% of the molecules.

Table 5.
Summary of the scaffold diversity of the eight databases analyzed in this work.

2.5.3 Molecular complexity and flexibility

The structural descriptors used to quantify fraction of sp^3 hybridized carbons (Fsp^3) [23, 58, 63, 70], fraction of chiral centers (CCF) [23, 59, 63, 70], fraction of aromatic atoms (Faro-atm), globularity [60], principal moments of inertia (PMI), normalized principal moments of inertia ratio (NRP) [61, 62], molecular complexity, shape index of Kier, and molecular flexibility were calculated with DataWarrior program [69] and MOE 2018.0101 [64]. **Figures 14–19** showed the descriptors utilized to evaluate the complexity and the molecular flexibility.

Tables 6–8 summarize the statistics of the distribution of Fsp^3, FCC, and Faro-atm of NPs and reference data sets. These results indicate that the NP data set has the largest complexity molecular in Fsp^3 (0.63) and CCF (0.16) and a low distribution of Faro-atm (0.67–0.78). In contrast, GNF, MMV, St. Jude, and GSK DBs are very similar in these three metrics with values between 0.25 and 0.37, 0.27 and 0.37, and 0.014 and 0.025, respectively. In contrast, the structural flexibility was evaluated with the index of form presenting all databases in the range of 0.41–0.58 indicating that many of the compounds present sphericity and intermediate molecular flexibility (data not presented).

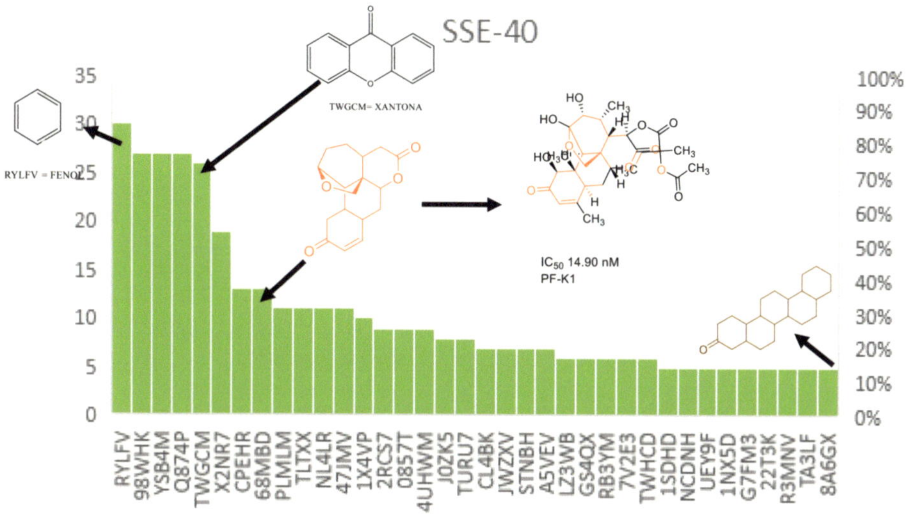

Figure 13.
Scaled Shannon entropy of the most frequent scaffolds with values ranging from 10 to 40 in natural products.

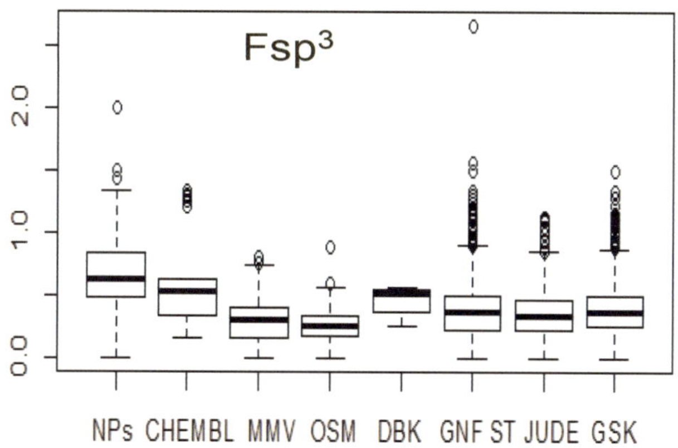

Figure 14.
Distribution of the fraction of sp³ hybridized carbons in different databases.

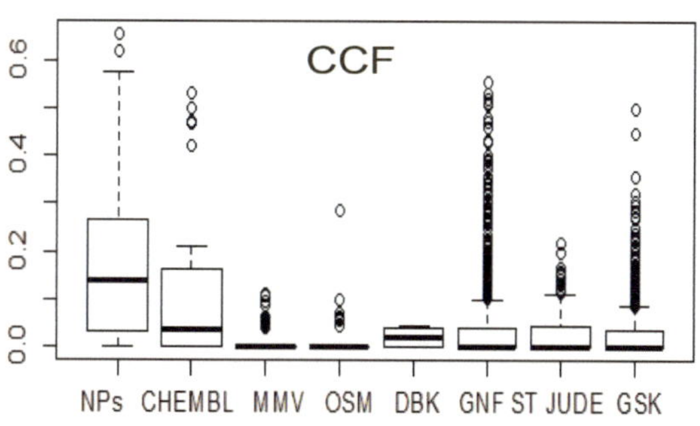

Figure 15.
Distribution of the fraction of chiral centers in different databases.

The descriptors globularity, PMI, and NRP did not prove to be suitable metrics to measure and differentiate the molecular complexity in the data sets evaluated. This is because the corresponding values computed for all data sets were very low

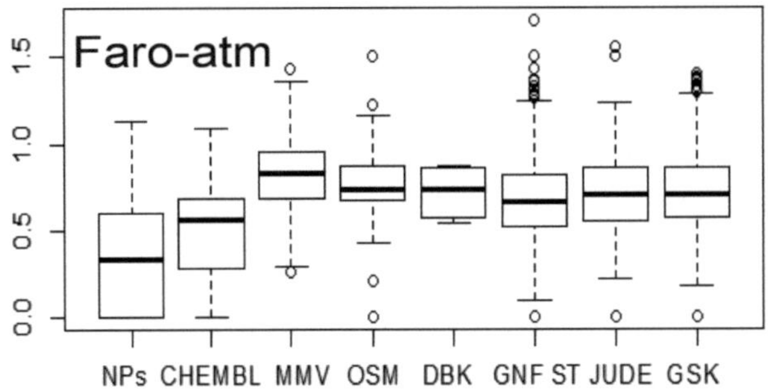

Figure 16.
Distribution of the fraction of aromatic atoms (Faro-atm) in different databases.

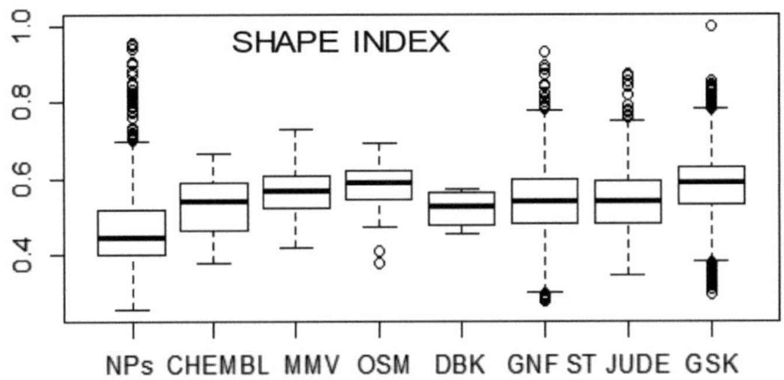

Figure 17.
Shape index distribution of different databases.

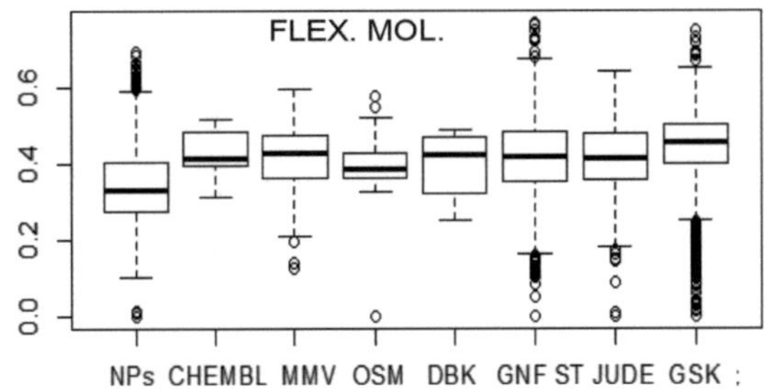

Figure 18.
Distribution of the molecular flexibility in different databases.

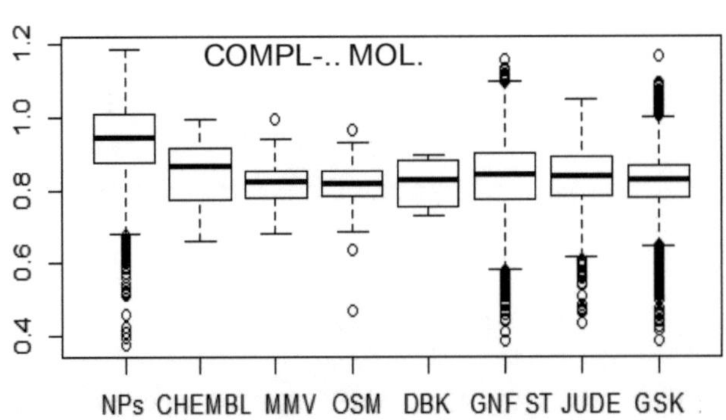

Figure 19.
Distribution of the molecular complexity in different databases.

Fraction of sp^3 hybridized atoms (Fsp^3)							
DBs	**Min**	**1qst**	**median**	**mean**	**3qrt**	**max**	**dev.st**
NPs	0.000	0.481	0.636	0.656	0.833	2.000	0.254
CHEMBL	0.167	0.342	0.536	0.621	0.627	1.333	0.374
MMV	0.000	0.167	0.300	0.316	0.402	0.800	0.190
OSM	0.000	0.174	0.255	0.277	0.338	0.893	0.145
DBK	0.250	0.438	0.519	0.463	0.545	0.565	0.175
GNF	0.000	0.227	0.364	0.377	0.500	2.667	0.207
STJUDE	0.000	0.222	0.333	0.353	0.471	1.136	0.178
GSK	0.000	0.250	0.375	0.372	0.500	1.500	0.180

Table 6.
Distribution of Fsp^3 in different databases.

Fraction of chiral centers (CCF)							
DBs	**min**	**1qst**	**median**	**mean**	**3qrt**	**max**	**dev.st**
NPs	0.000	0.033	0.139	0.161	0.267	0.656	0.145
CHEMBL	0.000	0.000	0.036	0.128	0.141	0.533	0.192
MMV	0.000	0.000	0.000	0.014	0.000	0.111	0.028
OSM	0.000	0.000	0.000	0.008	0.000	0.286	0.035
DBK	0.000	0.000	0.019	0.020	0.040	0.043	0.024
GNF	0.000	0.000	0.000	0.025	0.040	0.556	0.053
STJUDE	0.000	0.000	0.000	0.024	0.045	0.217	0.037
GSK	0.000	0.000	0.000	0.017	0.034	0.500	0.033

Table 7.
Distribution of FCC in different databases.

Fraction of aromatic atoms (Faro-atm)							
DBs	**min**	**1qst**	**median**	**mean**	**3qrt**	**max**	**dev.st**
NPs	0.000	0.000	0.324	0.341	0.600	1.133	0.294
CHEMBL	0.000	0.299	0.556	0.509	0.690	1.091	0.321
MMV	0.261	0.682	0.826	0.817	0.956	1.429	0.230
OSM	0.000	0.677	0.733	0.786	0.860	1.500	0.232
DBK	0.538	0.591	0.733	0.720	0.862	0.875	0.171
GNF	0.000	0.522	0.667	0.670	0.818	1.714	0.235
STJUDE	0.000	0.553	0.712	0.708	0.857	1.556	0.216
GSK	0.000	0.571	0.706	0.713	0.857	1.400	0.208

Table 8.
Distribution of fraction of aromatic atoms.

(close to zero) and did not differentiate the data sets (data not shown). The large molecular complexity of NPs measured is in agreement with previous studies using similar metrics [23, 63, 71].

3. Activity landscape modeling

The methods of modeling the landscape based on properties of the compounds (property landscape modeling (PLM)) is at the interface between experimental sciences and computational chemistry, being a frequent strategy to systematically describe the structure-property relationships (SPR) of the compound data set [72]. PLM have been used in medicinal chemistry in the stages of drug discovery with a quantitative, descriptive, and statistical approach to activity cliffs [72–74]. Structure-activity relationships (SARs), using the concept of modeling the activity landscape (activity landscape modeling ALM), are an increasing common practice in the drug discovery process to identify the activity cliffs, guide the optimization of compound hits, and to avoid the deleterious effects of the activity cliffs in the studies of the classic models of QSAR and in the search of structural similarity. In this

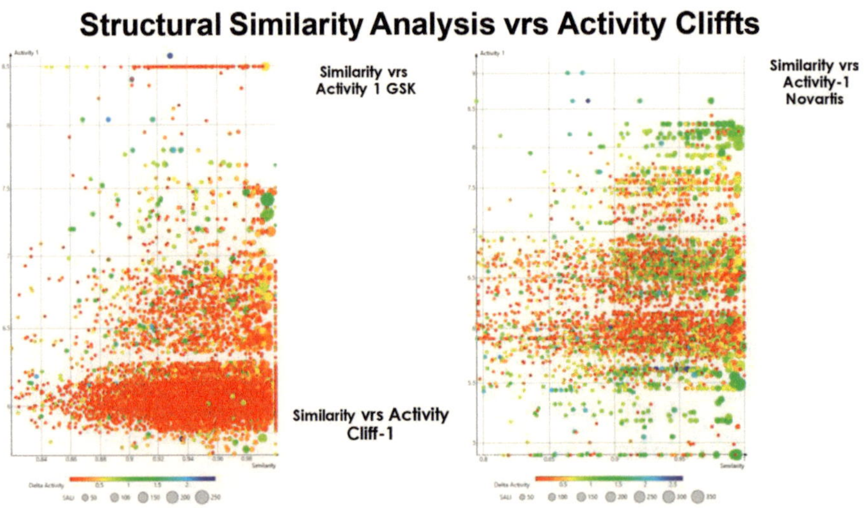

Figure 20.
Structural similarity compared with activity cliffs in NPAs.

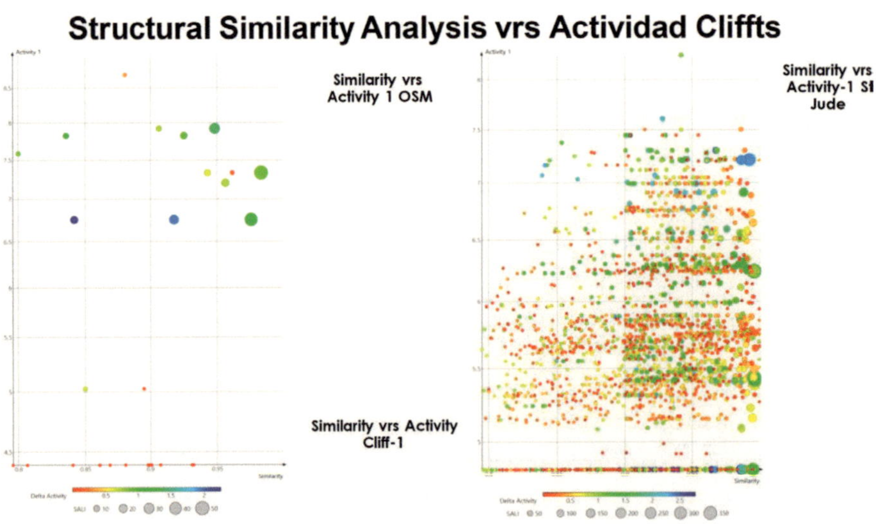

Figure 21.
Structural similarity compared with activity cliffs in GSK and Novartis (GNF).

research we analyze, through the web tool Activity Landscape Plotter (ALP) [72], a set of data from NPs from Panama with antimalarial activity against four strains of

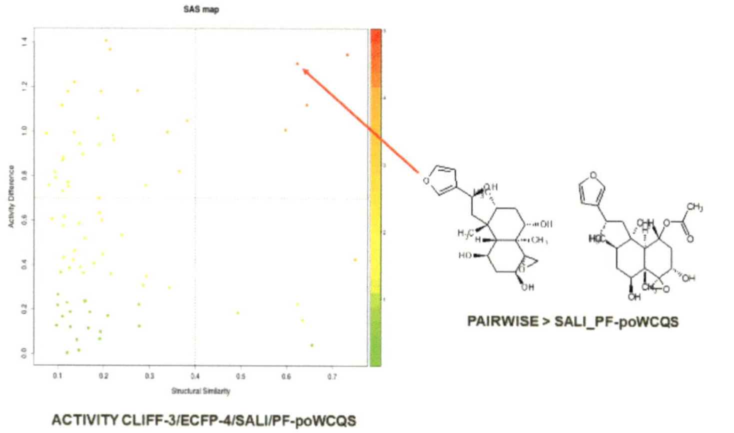

(a)

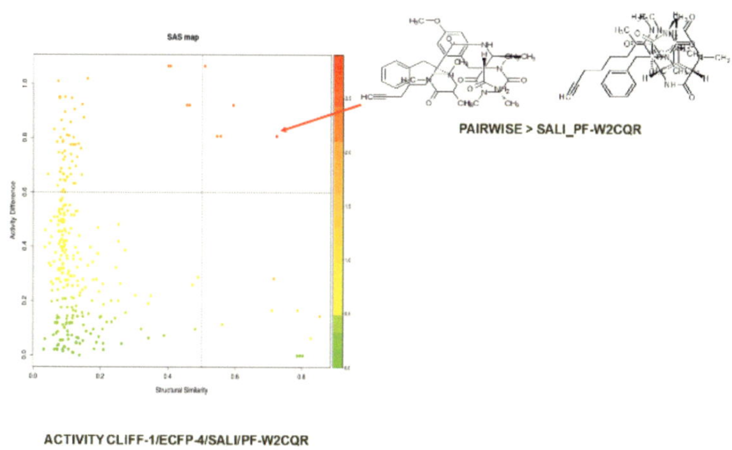

(b)

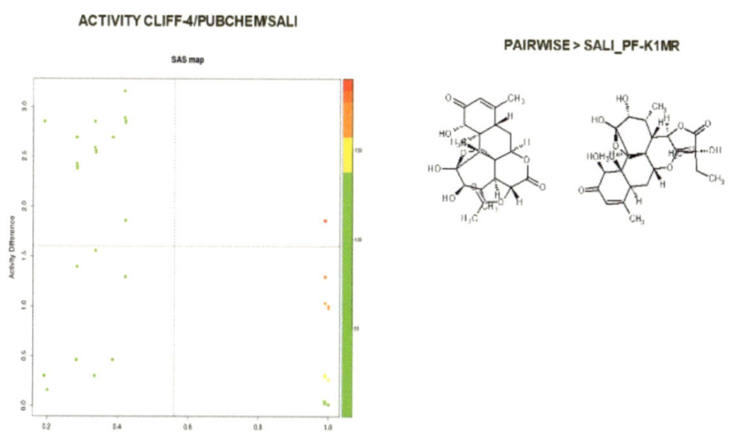

(c)

Figure 22.
SAS maps of compounds with antimalarial activity ((a), (b), and (c)) through the web tool activity landscape plotter.

Plasmodium falciparum in the erythrocyte gametocyte stage (**Figures 20** and **24**).

The generation and comparison of structure-activity pairs, by structure-activity similarity maps (SAS map). The SAS map has been used to link up structure and biological activity, based on a systematic pairwise comparison of all the compounds in a data set analyzed. We compare the values of structure-activity similarity, the activity difference, and structure-activity landscape index (SALI) to find the pairs of compounds with high molecular similarity and the activity difference that are located in the upper right quadrant of the SAS map (activity cliffs) [72–76]. **Figures 17–21** show SAS map in NP of Panama, NP published, GSK, and GNF. In SAS maps, data points are colored by density (**Figure 22**).

The SAS maps using the molecular fingerprints EFCP-4, MACCS keys, and PubChem led to the identification of a total of 26 pairs of compounds with structure-activity similarity ratios >0.50 and structure-activity landscape index values varying between 0.3 and 5.0. The web application Activity Landscape Plotter [72] is a tool that allows us to perform QSAR. The SAS generated represent 55 natural products isolated in Panama with antimalarial activity which were analyzed and compared the biological activities against strains of *Plasmodium falciparum* sensitive, resistant and multiresistant. The analysis with the parameters the (SAS / Tanimoto index / ECFP-4), a total of twenty-six pairs of compounds showed similarity values greater than 70%, sixteen pairs greater than 80% and only two pairs of compounds gave a similarity greater than 85%. While with activity cliffs, only three pairs of compounds show structural similarity correlated with the values of pIC50 activity [72, 77].

SAS maps are color-coded according to their intensity and we observe that most pairs of compounds with antimalarial activity show an intense red color. A nalyzed are located in the region of little structural similarity, indicating that the natural products have high structural diversity and low difference in activity, attributed to having similar functional groups in their molecules.

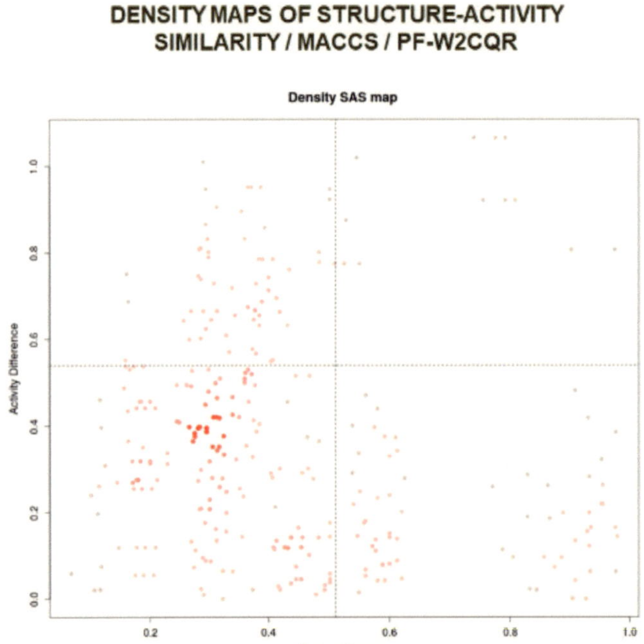

Figure 23.
DAS map with MACCS key fingerprint.

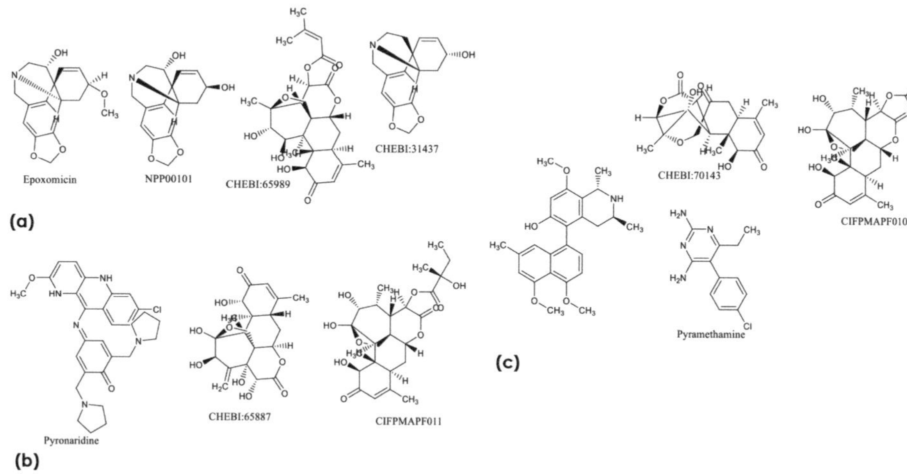

Figure 24.
Antimalarial compounds in NPs from Panama.

DAS maps represent the pairwise activity differences for each possible pair of compounds in an evaluated data set, against two biological targets. These maps permitted to differentiate if a structural modification can increase or decrease the activity under one target or other (**Figure 23**).

With this web application, we have carried out a QSAR study in a fast, simple, and easily interpretable way, obtaining three natural products as leading computational compounds for their optimization as ***Plasmodium falciparum*** blockers, which exhibit a gametocidal activity [78] (**Figure 24**).

4. Conclusion

The chemoinformatic analysis of the 20,364 compounds (1312 NPs and 19,052 synthetic (MMV, OSM, GNF, St. Jude, GSK, CHEMBL, and DrugBank)) indicates that so many natural products and synthetic products (S) share the same chemical space showing molecules that have similar structural properties. NPs present a greater diversity based on fingerprint than the synthetic compounds. Also, NPs have a higher proportion of chiral carbons and atoms with sp^3 hybridization and greater complexity, while synthetic products contain a greater proportion of aromatic atoms. Finally, concerning the properties related to cyclicity, relative shape, and flexibility, all have very similar values, which could explain the antimalarial activity of computationally determined compound hits in this work against *Plasmodium falciparum*-sensitive (3D7, D6, poW, D10) and chloroquine-resistant strains (W2, Dd).

Acknowledgements

The DAO acknowledges the SNI 2018 awards from SENACYT of Panama.

Conflict of interest

The authors declare that there are no financial or commercial conflicts of interest.

Author details

Dionisio A. Olmedo[1*] and José L. Medina-Franco[2]

1 CIFLORPAN Center for Pharmacognostic Research on Panamanian Flora, College of Pharmacy, University of Panama, Panama City, Panama

2 DIFACQUIM Research Group, Department of Pharmacy, School of Chemistry, National Autonomous University of Mexico (UNAM), Mexico City, Mexico

*Address all correspondence to: ciflorp4@up.ac.pa

References

[1] Newman DJ, Cragg GM. Natural products as sources of new drugs from 1981 to 2014. Journal of Natural Products. 2016;**9**:629-661. DOI: 10.1021/acs.jnatprod.5b01055

[2] Newman DJ, Cragg GM. Natural products as sources of new drugs over the 30 years from 1981 to 2010. Journal of Natural Products. 2016;**75**:311-335. DOI: 10.1021/np200906s

[3] Newman DJ, Cragg GM. Natural products as sources of new drugs over the last 25 years. Journal of Natural Products. 2007;**70**:461-477. DOI: 10.1021/np068054v

[4] Thomford NE, Senthebane DA, Rowe A, Munro D, Seele P, Maroyi A, et al. Natural products for drug discovery in the 21st century: Innovations for novel drug discovery. International Journal of Molecular Sciences. 2018;**19**:1578. DOI: 10.3390/ijms19061578

[5] Gurnani N, Mehta D, Gupta M, Mehta BK. Natural products: Source of potential drugs. African Journal of Basic & Applied Sciences. 2014;**6**:171-186. DOI: 10.5829/idosi.ajbas.2014.6.6.21983

[6] Hong J. Role of natural product diversity in chemical biology. Current Opinion in Chemical Biology. 2011;**15**:350-354. DOI: 10.1016/j.cbpa.2011.03.004

[7] Schreiber SL. Organic chemistry: Molecular diversity by design. Nature. 2009;**457**:153-154. DOI: 10.1038/457153a

[8] Schneider G, Grabowski K. Properties and architecture of drugs and natural products revisited. Current Chemical Biology. 2007;**1**:115-127. DOI: 10.2174/2212796810701010115

[9] Cragg GM, Newman DJ. Natural products: A continuing source of novel drug leads. Biochimica et Biophysica Acta. 2013;**1830**:3670-3695. DOI: 10.1016/j.bbagen.2013.02.008

[10] Sen S, Prabhu G, Bathula C, Hati S. Diversity-oriented asymmetric synthesis. Synthesis. 2014;**46**:2099-2121. DOI: 10.1055/s-0033-1341247

[11] van Hattum H, Waldmann H. Biology-oriented synthesis: Harnessing the power of evolution. Journal of the American Chemical Society. 2014;**136**:11853-11859. DOI: 10.1021/ja505861d

[12] Welsch ME, Snyder SA, Stockwell BR. Privileged scaffolds for library design and drug discovery. Current Opinion in Chemical Biology. 2010;**14**:347-361. DOI: 10.1016/j.cbpa

[13] Wetzel S, Bon RS, Kumar K, Waldmann H. Biology-oriented synthesis. Angewandte Chemie (International Ed. in English). 2011;**50**:10800-10826. DOI: 10.1002/anie.201007004

[14] Ertl P, Roggo R, Schuffenhauer A, Natural Product-likeness A. Score and its application for prioritization of compound libraries. Journal of Chemical Information and Modeling. 2008;**48**:68-74. DOI: 10.1021/ci700286x

[15] Mang C, Jakupovic S, Schunk S, Ambrosi H-D, Schwarz O, Jakupovic J. Natural products in combinatorial chemistry: An andrographolide-based library. Journal of Combinatorial Chemistry. 2006;**8**(2):268-274. DOI: 10.1021/cc050143n

[16] Wach JY, Gademann K. Reduce to the maximum: Truncated natural products as powerful modulators of biological processes. Synlett. 2012;**23**:163-170. DOI: 10.1055/s-0031-1290125

[17] Feher M, Schmidt JM. Property distribution: Differences between

drugs, natural products, and molecule from combinatorial chemistry. Journal of Chemical Information and Modeling. 2003;**43**:218-227. DOI: 10.1021/ci0200467

[18] Hert J, Irwin JJ, Laggner C, Keiser MJ, Shoichet BK. Quantifying biogenic bias in screening libraries. Nature Chemical Biology. 2009;**5**:479-483. DOI: 10.1038/nchembio.180

[19] Schenone M, Dancik V, Wagner BK, Clemons PA. Target identification and mechanism of action in chemical biology and drug discovery. Nature Chemical Biology. 2013;**9**:232-240. DOI: 10.1038/nchembio.1199

[20] Baell JB. Feeling nature's PAINS: Natural products, natural product drugs, and pan assay interference compounds (PAINS). Journal of Natural Products. 2016;**79**(3):616-628. DOI: 10.1021/acs.jnatprod.5b00947

[21] Egieyeh S, Syce J, Christoffels A, Malan SF. Exploration of scaffolds from natural products with antiplasmodial activities, currently registered antimalarial drugs and public malarial screen data. Molecules. 2016;**21**:104. DOI: 10.3390/molecules21010104

[22] Pilón-Jiménez B, Saldivar-Gonzalez F, Díaz-Eufracio BI, Medina-Franco JL. BIOFACQUIM: A mexican compound database of natural products. Biomolecules. 2019;**9**:31. DOI: 10.3390/biom9010031

[23] Olmedo DA, González-Medina M, Gupta MP, Medina-Franco JL. Cheminformatic characterization of natural products from Panama. Molecular Diversity. 2017;**21**(4):779-789. DOI: 10.1007/s11030-017-9781-4

[24] Pilon AC, Valli M, Dametto AC, MEF P, Freire RT, Castro-Gamboa I, et al. NuBBEDB: An updated database to uncover chemical and biological information from Brazilian biodiversity. Scientific Reports. 2017;7:7215-1-7215-12. DOI: 10.1038/s41598-017-07451-x

[25] Valli M, dos Santos RN, Figueira LD, Nakajima CH, Castro-Gamboa I, Andricopulo AD, et al. Development of a natural products database from the biodiversity of Brazil. Journal of Natural Products. 2013;**76**(3):439-444. DOI: 10.1021/np3006875

[26] Ntie-Kang F, Telukunta KK, Döring K, Simoben CV, Moumbock AF, Malange YI, et al. NANPDB: A resource for natural products from northern african sources. Journal of Natural Products. 2017;**80**(7):2067-2076. DOI: 10.1021/acs.jnatprod.7b00283

[27] Chen CY-C. TCM database@Taiwan: The world's largest traditional chinese medicine database for drug screening in silico. PLoS ONE. 2011;**6**(1):e15939. DOI: 10.1371/journal.pone.0015939

[28] Ye H, Ye L, Kang H, Zhang D, Tao L, Tang K, et al. HIT: Linking herbal active ingredients to targets. Nucleic Acids Research. 2011;**39**:D1055-D1059. DOI: 10.1093/nar/gkq1165

[29] Mangal M, Sagar P, Singh H, Raghava GPS, Agarwal SM. NPACT: Naturally occurring plant-based anti-cancer compound-activity-target database. Nucleic Acids Research. 2013;**41**:D1124-D1129. DOI: 10.1093/nar/gks1047

[30] Bhalerao SA, Verna DR, D'Souza LR, Teli NC, Didwana VS. Chemoinformatics: The application of informatic methods to solve chemical problem. Research Journal of Pharmaceutical, Biological and Chemical Sciences. 2007;**4**(3):475-499

[31] Wenderski TA, Stratton CF, Bauer RA, Kopp F, Tan DS. Principal component analysis as a tool for library design: A case study investigating natural products, brand-name drugs, natural product-like libraries,

and drug-like libraries. Methods in Molecular Biology. 2015;**1263**:225-242. DOI: 10.1007/978-1-4939-2269-7_18

[32] Medina-Franco JL, Mayorga-Martínez K, Giulianotti MA, Houghten RA, Pinilla C. Visualization of the chemical space in drug discovery. Current Computer-Aided Drug Design. 2008;**4**:322-333. DOI: 10.2174/157340908786786010

[33] Osolodkin DI, Radchenko EV, Orlov AA, Voronkov AE, Palyulin VA, Zefirov NS. Progress in visual representations of chemical space. Expert Opinion on Drug Discovery. 2015;**10**(9):959-973. DOI: 10.1517/17460441

[34] Ringner M. What is principal component analysis? Nature Biotechnology. 2018;**26**:303-304. DOI: 10.1038/nbt0308-303

[35] Jolliffe I. Principal component analysis. In: Everitt BS, Howell DC, editors. Encyclopedia of Statistics in Behavioral Science. Vol. 3. Aberdeen, Chichester, UK: University of Aberdeen, John Wiley and Sons, Ltd; 2005. pp. 1580-1584. DOI: 10.1002/0470013192.bsa501

[36] Clemons PA, Wilson JA, Dančík V, Muller S, Carrinski HA, Wagner BK, et al. Quantifying structure and performance diversity for sets of small molecules comprising small-molecule screening collections. Proceedings of the National Academy of Sciences of the United States of America. 2011;**108**:6817-6822. DOI: 10.1073/pnas.1015024108

[37] Djuric SW, Akritopoulou-Zanze I, Cox PB, Galasinski S. Compound collection enhancement and paradigms for high-throughput screening-an update. Annual Reports in Medicinal Chemistry. 2010;**45**:409-428

[38] Rogers D, Hahn M. Extended-connectivity fingerprints. Journal of Chemical Information and Modeling. 2010;**50**:742-754. DOI: 10.1021/ci100050t

[39] Medina-Franco JL, Martínez-Mayorga K, Peppard TL, Del Rio A. Chemoinformatic analysis of GRAS (generally recognized as safe) flavor chemicals and natural products. PLoS One. 2012;7(11):e50798. DOI: 10.1371/journal.pone.0050798

[40] Durant JL, Leland BA, Henry DR, Nourse JG. Reoptimization of MDL keys for use in drug discovery. Journal of Chemical Information and Computer Sciences. 2002;**42**(6):1273-1280

[41] Bolton EE, Wang Y, Thiessen PA, Bryant SH. PubChem: Integrated platform of small molecules and biological activities. Annual Reports in Computational Chemistry. 2008;**31**(4):217-241

[42] Rogers D, Brown R, Hahn M. Using extended-connectivity fingerprints with Laplacian-modified Bayesian analysis in high-throughput screening. Journal of Biomolecular Screening. 2005;**10**:682-686. DOI: 10.1177/1087057105281365

[43] Hert J, Willet P, Wilton DJ, Acklin P, Azzaoui K, Jacoby E, et al. Comparison of topological descriptors for similarity-based virtual screening using multiple bioactive reference structures. Organic and Biomolecular Chemistry. 2004;**2**:3256-3266. DOI: 10.1039/B409865J

[44] Barnard JM, Downs GM. Chemical fragment generation and clustering software. Journal of Chemical Information and Computer Sciences. 1997;**37**:141-142. DOI: 10.1021/ci960090k

[45] González-Medina M, Prieto-Martínez FD, Owen JR, Medina-Franco JL. Consensus diversity plots: A global diversity analysis of chemical libraries. Journal of Cheminformatics. 2016;**8**:63. DOI: 10.1186/s13321-016-0176-9

[46] Bajusz D, Rácz A, Károly Héberger K. Why is Tanimoto index an appropriate choice for fingerprint-based similarity calculations? Journal

of Cheminformatics. 2015;7:20. DOI: 10.1186/s13321-015-0069-3

[47] Owen JR, Nabney IT, Medina-Franco JL, López-Vallejo F. Visualization of molecular fingerprints. Journal of Chemical Information and Modeling. 2011;**51**:1552-1563. DOI: d10.1021/ci1004042

[48] Skinnider MA, Dejong CA, Franczak BC, McNicholas PD, Nathan A, Magarvey NA. Comparative analysis of chemical similarity methods for modular natural products with a hypothetical structure enumeration algorithm. Journal of Cheminformatics. 2017;**9**:46. DOI: 10.1186/s13321-017-0234-y

[49] Medina-Franco JL. Discovery and development of lead compounds from natural sources using computational approaches. In: Mukherjee P, editor. Evidence-Based Validation of Herbal Medicine. Amsterdam, The Netherlands: Elsevier; 2015. pp. 455-475

[50] Xu Y-J, Johnson M. Using molecular equivalence numbers to visually explore structural features that distinguish chemical libraries. Journal of Chemical Information and Computer Sciences. 2002;**42**:912-926. DOI: 10.1021/ci025535l

[51] Xu Y-J, Johnson M. Algorithm for naming molecular equivalence classes represented by labeled pseudographs. Journal of Chemical Information and Computer Sciences. 2001;**41**:181-185. DOI: 10.1021/ci0003911

[52] Lopez-Vallejo F, Castillo R, Yepez-Mulia L, Medina-Franco JL. Benzotriazoles and indazoles are scaffolds with biological activity against Entamoeba histolytica. Journal of Biomolecular Screening. 2011;**16**:862-868. DOI: 10.1177/1087057111414902

[53] Hu Y, Bajorath J. Quantifying the tendency of therapeutic target proteins to bind promiscuous or selective compounds. PLoS ONE. 2015;**10**:e0126838. DOI: 10.1371/journal.pone.0126838

[54] Lipkus AH, Yuan Q, Lucas KA, Funk SA, Bartelt WF 3rd, Schenck RJ, et al. Structural diversity of organic chemistry. A scaffold analysis of the CAS registry. The Journal of Organic Chemistry. 2008;**73**:4443-4451. DOI: 10.1021/jo8001276

[55] Krier M, Bret G, Rognan D. Assessing the scaffold diversity of screening libraries. Journal of Chemical Information and Modeling. 2006;**46**:512-524. DOI: 10.1021/ci050352v

[56] Medina-Franco JL, Martínez-Mayorga K, Bender A, Scior T. Scaffold diversity analysis of compound data sets using an entropy-based measure. QSAR and Combinatorial Science. 2009;**28**:1551-1560. DOI: 10.1002/qsar.200960069

[57] Yongye AB, Waddell J, Medina-Franco JL. Molecular scaffold analysis of natural products databases in the public domain. Chemical Biology & Drug Design. 2012;**80**:717-724. DOI: 10.1111/cbdd.12011

[58] Todeschini R, Consonni V. Molecular Descriptors for Chemoinformatics. Germany: Wiley-VCH; 2009

[59] Clemons PA, Bodycombe NE, Carrinski HA, Wilson JA, Shamji AF, Wagner BK, et al. Small molecules of different origins have distinct distributions of structural complexity that correlate with protein-binding profiles. Proceedings of the National Academy of Sciences of the United States of America. 2010;**107**:18787-18792. DOI: 10.1073/pnas.1012741107

[60] Meyer AY. Molecular mechanics and molecular shape. III. Surface area and cross-sectional areas of organic molecules. Journal of Computational Chemistry. 1986;7:144-152. DOI: 10.1002/jcc.540070207

[61] Sauer WHB, Schwarz MK. Molecular shape diversity of combinatorial libraries: A prerequisite for broad bioactivity. Journal of Chemical Information and Computer Sciences. 2003;**43**:987-1003. DOI: 10.1021/ci025599w

[62] Rolfe A, Lushington GH, Hanson PR. Reagent based DOS: A 'click, click, cyclize strategy to probe chemical space. Organic & Biomolecular Chemistry. 2010;**8**:2198-2203. DOI: 10.1039/b927161a

[63] Méndez-Lucio O, Medina-Franco JL. The many roles of molecular complexity in drug discovery. Drug Discovery Today. 2017;**22**:120-126. DOI: 10.1016/j.drudis.2016.08.009

[64] Molecular Operating Environment (MOE). 2018.0101. 910-1010 Sherbrooke, St. W. Montreal, QC H3A 2R7; Canada: Chemical Computing Group, Corporate Headquarters Montreal

[65] Lipinski CA, Lombardo F, Dominy BW, Feeney PJ. Experimental and computational approaches to estimate solubility and permeability in drug discovery and development settings. Advanced Drug Delivery Reviews. 2001;**46**(1-3):3-26. DOI: 10.1016/S0169-409X(00)00129-0

[66] Veber DF, Johnson SR, Cheng HY, Smith BR, Ward KW, Kopple KD. Molecular properties that influence the oral bioavailability of drug candidates. Journal of Medicinal Chemistry. 2002;**45**(12):2615-2623. DOI: 10.1021/jm020017n

[67] Leeson PD, Springthorpe B. The influence of drug-like concepts on decision-making in medicinal chemistry. Nature Reviews. Drug Discovery. 2007;**6**(11):881-890. DOI: 10.1038/nrd2445

[68] RStudio Team. RStudio: Integrated Development for R. Boston: RStudio, Inc.; 2015. Available form: http://www.rstudio.com/

[69] Sander T, Freyss J, Von Korff M, Rufener C. Datawarrior: An open-source program for chemistry aware data visualization and analysis. Journal of Chemical Information and Modeling. 2015;**55**:460-473. DOI: 10.1021/ci500588j

[70] Godden JW, Bajorath J. Analysis of chemical information content using Shannon entropy. In: Lipkowitz KB, Cundari TR., editors. Reviews in Computational Chemistry. Hoboken: John Wiley & Sons, Inc.; 2007. pp. 263-289. DOI: 10.1002/9780470116449.ch5

[71] Lovering F, Bikker J, Humblet C. Escape from flatland: Increasing saturation as an approach to improving clinical success. Journal of Medicinal Chemistry. 2009;**52**:6752-6756. DOI: 10.1021/jm901241

[72] González-Medina M, Prieto-Martínez FD, Naveja J, Méndez-Lucio O, El-Elimat T, Pearce CJ, et al. Chemoinformatic expedition of the chemical space of fungal products. Future Medicinal Chemistry. 2016;**8**:1399-1412. DOI: 10.4155/fmc-2016-0079

[73] González-Medina M, Méndez-Lucio O, Medina-Franco JL. Activity landscape plotter: A web-based application for the analysis of structure–activity relationships. Journal of Chemical Information and Modeling. 2017;**57**:397-402

[74] Cruz-Monteagudo M, Medina-Franco JL, Pérez-Castillo Y, Nicolotti O, Cordeiro MNDS, Borges F. Activity cliffs in drug discovery: Dr Jekyll or Mr Hyde? Drug Discovery Today. 2014;**19**(8):1069-1080

[75] Bajorath J. Modeling of activity landscapes for drug discovery. Expert Opinion on Drug Discovery. 2012;**7**(6):463-473

[76] Garcia-Sanchez MO, Cruz-Monteagudo M, Medina-Franco JL. Challenges and advances in computational chemistry on physics quantitative structure-epigenetic activity relationship. In: Lezcznski J, Roy K, editors. Advances in QSAR Modeling: Application in Pharmaceutical, Chemical, Foods, Agricultural and Environmental Science, 24. Gewerbestrasse 11, 6330 Cham, Switzerland: Springer Nature. Springer International Publishing AG; 2017. pp. 303-338

[77] Medina-Franco JL. Scanning structure–activity relationships with structure–activity similarity and related maps: From consensus activity cliffs to selectivity switches. Journal of Chemical Information and Modeling. 2012;**52**(10):2485-2493. DOI: 10.1021/ci300362x

[78] Kiszewski AE. Blocking plasmodium falciparum malaria transmission with drugs: The gametocytocidal and sporontocidal properties of current and prospective antimalarials. Pharmaceuticals. 2011;**4**(1):44-68. DOI: 10.3390/ph4010044

Section 2

Drug Design and Develpment by Chemical Tools

Chapter 7

Prologue: Cheminformatics and Its Applications

Azhar Rasul

1. Introduction

Cheminformatics is a field of information technology that uses informational and computational techniques to provide a deeper understanding and solutions of problems of chemistry. Cheminformatics strategies originally emerged as vehicle in drug discovery where large libraries of compounds are evaluated for specific functionality or therapeutic effects [1].

Drug discovery is a highly systematic multistep procedure for the identification of new medicines [2]. Chemical toolsets including chemical probes, RNAi, and chemoproteomics have helped scientists to identify and validate novel druggable targets for therapeutic interventions [3–6]. Target validation is of pivotal importance in determining the suitability of a new target for further clinical evaluation. Following the process of target validation, hit identification and lead discovery process involves establishment of high throughput screening (HTS) systems as well as development of chemical tool compound libraries [2]. The next critical phase of the drug discovery process is pharmacokinetics and pharmacodynamic profiling of lead compounds [7] and investigation of Absorption, Distribution, Metabolism, Excretion and Toxicity (ADMET) properties [8]. Various critical steps in drug discovery involve the applications of cheminformatics such as compound selection, virtual library generation, in silico-based screening, HTS, HTS data mining, and in silico ADMET profiling [9].

In addition, cheminformatics have also helped the scientists to develop and optimize delivery of molecules to intracellular targets for therapeutic implications [10], thus, provided solutions for various unmet medical needs.

Conjugation of therapeutic entities with peptide delivery molecules, especially cell-penetrating peptides (CPPs), has the potential to increase the therapeutic efficacy by enhancing the ability of therapeutics to reach specific intracellular targets [11]. Preclinical evaluations of CPP-mediated therapeutics have shown promising results in disease models that also prompted clinical trials in some cases. These outcomes have, thus, opened new perspectives for CPPs in the development of well-tolerated and specifically targeted human therapies [12]. Thus, insights into current approaches and potential of CPP-based drug delivery systems are presented for greater understanding of readers about powerful promises and clinical efficacy of CPP-based therapeutics.

Cheminformatics and its applications presents the applications of two fields, chemical biology and bioinformatics, in drug discovery, thus, providing comprehensive description of modern technologies such as structure-based drug design, molecular docking, high throughput screening, and pharmaceutical profiling, which are all critical steps for the development of successful marketable drugs. With the invention of advanced and modern techniques in bioinformatics, the process of drug discovery has become faster and economical. Bioinformatics-based computational techniques have provided platform for large-scale screening of small molecules and chemical biology has served pharmaceutics for the validation of obtained data from computer-aided techniques, thus, both of the fields go hand in hand to revolutionize the field of drug discovery. Keeping in view of the emerging trends on cheminformatics in drug discovery, this book is designed to enable scientists to understand the fundamentals of drug discovery. Beginning with the highlights of the historical timeline of drug discovery, this book simply and succinctly educates its readers about screening methods, medicinal chemistry strategies in drug design, lead generation, testing the bioactivity of leads, lead optimization, clinical trial basics, as well as challenges of drug discovery such as cell-penetrating peptides and acceleration of chemical tool discovery by academic collaborations. This book will provide a clearer picture of cheminformatics and its applications and will be useful for scientific community working in the arena of drug discovery. Several recent developments are also overviewed, which will make it valuable for academicians and scientists.

Author details

Azhar Rasul
Department of Zoology, Government College University Faisalabad, Pakistan

*Address all correspondence to: drazharrasul@gmail.com

References

[1] Wishart DS. Introduction to cheminformatics. Current Protocols in Bioinformatics. 2016;**53**:14 11 11-14 11 21

[2] Hughes JP, Rees S, Kalindjian SB, Philpott KL. Principles of early drug discovery. British Journal of Pharmacology. 2011;**162**:1239-1249

[3] Kurreck J. Expediting target identification and validation through RNAi. Expert Opinion on Biological Therapy. 2004;**4**:427-429

[4] Bantscheff M, Drewes G. Chemoproteomic approaches to drug target identification and drug profiling. Bioorganic & Medicinal Chemistry. 2012;**20**:1973-1978

[5] Bunnage ME, Chekler EL, Jones LH. Target validation using chemical probes. Nature Chemical Biology. 2013;**9**:195-199

[6] Moustakim M, Felce SL, Zaarour N, Farnie G, McCann FE, Brennan PE. Target identification using chemical probes. Methods in Enzymology. 2018;**610**:27-58

[7] Tuntland T, Ethell B, Kosaka T, Blasco F, Zang RX, Jain M, et al. Implementation of pharmacokinetic and pharmacodynamic strategies in early research phases of drug discovery and development at Novartis Institute of Biomedical Research. Frontiers in Pharmacology. 2014;**5**:174

[8] Wang J, Skolnik S. Recent advances in physicochemical and ADMET profiling in drug discovery. Chemistry & Biodiversity. 2009;**6**:1887-1899

[9] Xu J, Hagler A. Chemoinformatics and drug discovery. Molecules. 2002;**7**:566-600

[10] Gautam A, Chaudhary K, Kumar R, Sharma A, Kapoor P, Tyagi A, et al. In silico approaches for designing highly effective cell penetrating peptides. Journal of Translational Medicine. 2013;**11**:74

[11] Derakhshankhah H, Jafari S. Cell penetrating peptides: A concise review with emphasis on biomedical applications. Biomedicine & Pharmacotherapy. 2018;**108**:1090-1096

[12] Guidotti G, Brambilla L, Rossi D. Cell-penetrating peptides: From basic research to clinics. Trends in Pharmacological Sciences. 2017;**38**:406-424

Chapter 8

Accelerating Chemical Tool Discovery by Academic Collaborative Models

Bahne Stechmann and Wolfgang Fecke

Abstract

The development of chemical tool compounds becomes increasingly important for chemical biology research projects in many disciplines of life sciences. In addition, they form essential parts in both academic and industrial drug discovery efforts. The required expertise and technology platforms for the identification and optimization of these potent and target-selective small molecules often exceed the capabilities of academic groups and smaller companies. Over the years, several initiatives were created all over the world which address this issue by either creating networks or consortia of academic institutes, public-private partnerships with industry, or even dedicated new research infrastructures for chemical biology. Several of these organizations and their different access models will be described. We will focus in particular on the model of EU-OPENSCREEN ERIC, a new European Research Infrastructure which was founded in 2018 and consists of more than 20 partner institutes from eight countries.

Keywords: academic consortia, chemical tools, drug, pharmacological screening

1. Introduction

In the last decade, the interdisciplinary field of chemical biology has emerged from the need to better understand the role of proteins or signaling pathways in cellular systems and whole organisms than it was previously feasible with more classical genetic tools or methods. Rather than changing the levels of proteins, or blocking completely their expression or activity, by deleting or overexpressing their respective DNA or RNA sequences, it is now becoming more and more possible to precisely modulate their function in a time- and concentration-dependent manner using potent, selective and cell-permeable chemical compounds. Although the relevance of these so-called chemical tool compounds or probes for solving basic mechanistical questions in life sciences is indisputable [1], their role often extends into the fields of pharmacology and molecular medicine. In fact, chemical tools are playing an important role in the validation of newly identified drug targets in pharmaceutical companies, and might even serve as starting points for the development of new therapeutics.

Despite recent technological advances in areas such as cryo-electron microscopy (Cryo-EM) [2], the major approach for identifying bioactive substances is still the systematic testing of compound collections, often comprising many thousands or

 IntechOpen

even millions of individual substances, with target- or pathway-specific biological assays which are designed to produce reproducible biological activities with high signal-to-noise ratios under experimental conditions which are fast, miniaturized and therefore cost-effective [3]. This approach is technically and logistically challenging and, in the past, could only be performed by large pharmaceutical companies. In addition to experienced personnel, it requires large facilities with often expensive equipment for compound storage, automated liquid handling and sensitive detection of biological reactions. In recent years, however, this picture started to change. In the wake of the sequencing of the human genome, mostly larger academic institutions started to create their own screening and translational drug discovery centers because many new potential drug targets were suddenly becoming available for which a solid understanding of their physiological roles and molecular mechanisms were missing. At the same time, pharmaceutical companies faced increased pressures due to high drug development costs, often resulting in down-sized research budgets and cost cutting exercises combined with a general trend of becoming risk-averse towards innovative drug targets with potential high failure rates [4]. As a result, many experienced industrial 'drug hunters' found employment in academic chemical probe discovery centers, supporting their efforts and helping to alleviate some of the initial issues these centers faced [5].

In this chapter we describe some of these new initiatives which were created to develop chemical tool compounds outside of the traditional pharmaceutical industry, highlighting their particular strengths, challenges and access models for the mostly academic scientific community.

2. Developing chemical probes in academic networks

At the beginning of the 21st century, academic institutions first began to implement dedicated assay development and screening centers which were soon followed by reports on the systematic testing of small molecule compound libraries in the US [6]. Comparable efforts in Europe's research institutes immediately received much attention, fostering collaborations between chemistry and biology groups and the establishment of academic screening platforms of diverse size. However, single platforms alone could not support comprehensively the needs of academic or industrial users due to limited chemical diversity of their compound collections and/or limited technical capabilities, and big pharma platforms were at that time not open to academic users. Pooling and coordination of public resources and expertise became imperative. Therefore, the efforts in the US were replicated with similar initiatives in countries such as France (Chimiothèque Nationale), Germany (ChemBioNet), Spain (ChemBioBank) and several others. Some years later, long-term cooperations between academic centers from different countries as well as public-private partnerships were established. We will describe some of these initiatives in more detail and will put particular emphasis on the collaborative model of the youngest organization for chemical biology, the EU-OPENSCREEN ERIC.

2.1 The molecular libraries program (MLP) in the US

The large, NIH-funded MLP was created in 2004 with the ambitious goal of creating a small molecule probe for every human protein in order to define the functions of genes, cells, and whole organisms in health and disease. The three components of the initiative were essentially: (a) a network of comprehensive and specialized screening centers plus specialized chemistry centers, (b) several cheminformatics approaches which included also a newly created public compound database called PubChem with assay metadata, and (c) initiatives to generally advance technologies in the fields of chemical diversity, assay development and screening [6]. The aim was always to publish the new chemical probes and associated data immediately so that compounds could be used by the academic scientific community not only for basic research questions, but also for mechanistic validation of potentially disease-relevant drug targets and drug development.

Individual scientists could apply for funding to the NIH for their assay development and screening projects. Successfully, peer-reviewed projects were taken on board by one of the MLP centers, and high-throughput screens were conducted with a library which, by the end of the program, consisted of 390.000 compounds. About 5% of these molecules with often novel scaffolds were delivered by the academic synthetic chemistry community. In many cases, further chemical optimization yielded probes against protein targets which were deemed challenging or even 'undruggable'. Overall, during the 10-year period of the program, a total of 375 chemical tool compounds were developed against a broad range of target classes. 18 of these compounds were considered sufficiently interesting to serve as starting points for the development of therapeutics against a total of eight disease targets or target classes, and were licensed to biotech and pharmaceutical companies [7]. In light of the investment into the MLP it is debatable whether the ratio of probes to drug candidates can be regarded as a success or a disappointment but it certainly highlights the difficulties that chemical biologists are facing when they want to keep up with the speed of biological discoveries while translating academic findings into therapeutics.

2.2 The chemical biology consortium Sweden (CBCS)

Although much smaller than the MLP in the US, this example of a national consortium can highlight very well the particular strengths of a focused organization with only a few members. CBCS, with two nodes at the Karolinska Institutet and Umeå University, was founded as a non-for-profit research infrastructure for chemical biology in 2010 [8] by researchers from Biovitrum (former Pharmacia and Upjohn) and became an integrated platform of SciLifeLab, an already existing national centre for molecular biosciences, in 2013 [9]. The combined platform can investigate both chemical and genetic perturbations in biological systems. CBCS wants to enable high level basic research with open access publications while at the same time linking up academic and industrial groups. Complementary to CBCS, SciLifeLab offers a dedicated platform for drug discovery and development, with the clear goal of accelerating projects with translational potential. After nearly 10 years of operation, the consortium has produced more than 130 co-authored publications and 11 patent applications while scientific data provided the basis of six start-up companies [10].

Users are encouraged to discuss in more detail project proposals with the CBCS staff prior to the submission of the official application. A proposal template, user agreements and estimated costs of typical screening and chemistry projects are available online. Project proposals are evaluated by an independent 'Project Review Committee' (PRC), which meets biannually. Prioritized projects may be subsidized, with the remaining costs covered by the applicant. Implemented projects are periodically re-evaluated by the Project Review Committee as they progress to pre-defined milestones. A project plan for a so-called "large collaborative project" may run over a maximum of 2 years for which the user is expected to cover the costs for all reagents and consumables, including a compound access fee for plating of library compounds. There are also "small collaborative projects" which involve only limited CBCS support for maximal 2 weeks, e.g. a short-term access to a specialized instrument such as an imaging plate reader. For these projects, no PRC application is required but they are undertaken with a "first come—first served" policy based on available resources [10]. In line with the open access data policy of the CBCS, the applicant and the CBCS agree upon a clear publication strategy before the implementation of the project. The target user group of CBCS are academic researchers at Swedish research institutions, who aim to develop chemical probes on a collaborative basis.

It is worth looking in more detail into the services CBCS can offer to their academic customers. The consortium assists in assay development for both biochemical and cell-based assays, gives access to the SciLifeLab compound collection and provides medicinal and computational chemistry expertise for hit validation and optimization. This model is very similar to the service offerings of the much larger European research infrastructure EU-OPENSCREEN which is being discussed below. In addition, mechanism-of-action studies can be performed with often specialized technologies such as cellular thermal shift assays (CETSA) [11]. In fact, the development of CETSA is a good example on how an expert consortium such as CBCS can impact and further develop disrupting technologies in collaboration with local academic groups and commercial partners (here: Pelago Biosciences). Starting life as a low throughput assay, CETSA is now amenable to high throughput screening [12]. Scientists usually come to the CBCS with the concept for a biological assay and first experimental data. They have then the chance to work further on the assay in the CBCS laboratories under guidance of their expert scientists, enabling in parallel scientific services and the education of users [10].

The CBCS compound collection consists of more than 200.000 compounds with high chemical diversity which are routinely quality controlled. While many of these compounds were donated by the pharmaceutical company Biovitrum, the library was further expanded with sets from commercial vendors and donations by other biotech companies. Importantly, the strategy has always been to build a modular collection of sub-libraries which can be adapted to the needs of each academic screening project, based mainly on assay throughput and cost per data point. For instance, in addition to a diverse primary screening set of 35.000 compounds, there are also focused libraries for particular target classes such as kinases, G-protein coupled receptors, agrochemicals etc., as well as a set of approved drugs [10]. This is very different to the concept of EU-OPENSCREEN which offers a high throughput screening set of 100.000 commercial compounds to their users, with the goal to have that set screened in almost all projects so that each compound becomes associated with "positive" and "negative" screening data from as many projects as possible (see below).

Overall, between 2010 and 2018 more than 400 collaborative projects with 236 individual users in Sweden were discussed. User interest grew continuously during these 8 years, currently leading to approximately one new project discussion per week. About 25% of discussions result in large project PRC applications while others obtain small project limited support, all documented in, on average, 20 publications per year [10].

2.3 Public private initiatives: SGC and ELF

In industry, chemical tool compounds play an important role as pharmaceutical modulators of novel drug targets. Typically, they are being used for testing a particular disease hypothesis and for validating the chemical tractability of newly discovered candidate proteins or signaling pathways for which otherwise comparatively little information is available. Sometimes their properties are even sufficient to act as starting points for drug discovery programs. The development of compounds with required potency and, most importantly, selectivity towards individual members of a protein class can be a formidable task even for larger pharmaceutical or biotech companies. It came therefore as no surprise that in 2009 several industrial partners decided to collaborate in a pre-competitive manner and initiated a public-private partnership (PPP) with leading academic institutes in the field of chemical biology. The aim was to develop high-quality chemical tool compounds for families of understudied proteins of potential therapeutic value, for instance epigenetic and other transcriptional modulators.

The chosen academic partners in that particular PPP were the universities of Oxford and Toronto which had already formed the so-called Structural Genomics Consortium (SGC) in 2004 with the goal of determining the three-dimensional structures of proteins with therapeutic relevance. The SGC advocates open access partnerships between industry and academia and is committed to make their chemical tool compounds available without any restrictions. In the last 10 years, and with financial support by several pharmaceutical companies, more than 50 chemical probes in the areas of epigenetics and kinase signaling were developed [13, 14]. Furthermore, seven pharmaceutical companies made their chemical tool compounds from older research programs available to the scientific community, including protocols, controls and associated data [15]. Efforts are now underway, under the umbrella of the Innovative Medicines Initiative (IMI), to expand the initial collection of compounds further by focusing not only on the protein classes which were selected in the past but also on the development of new technologies, making the identification and profiling of tool compounds generally faster and more cost-effective [16].

Another PPP initiative supported by the IMI is the European Lead Factory (ELF) [17] which is a consortium of 20 partners, currently among them the universities of Oxford and Dundee while several other universities, research organizations and companies in the UK, Netherlands and Germany were former partners. The project was launched in 2013 and came to an end in 2018, with a follow-up five-year project funded in the same year [18]. During its lifetime, the ELF established a selection of about 550.000 compounds which are generally not commercially available. 300.000 of these were donated by seven participating pharmaceutical companies, while the rest was synthesized by medicinal chemistry partner companies during the last 5 years. Both the compound management facility in the UK and the high throughput screening center in the Netherlands were formerly part of pharmaceutical companies and able to perform screening operations and chemistry services such as hit optimization and modeling according to industry standards. The Oxford Biotechnology group of the SGC was selected as a key contributor of 3D co-crystal structures which are essential for compound optimization. During the lifetime of the project, more than 80 drug discovery programs across most therapeutic areas were pursued. By March 2018, two partnering deals between the respective project owner and one of the pharmaceutical company partners had emerged. Importantly, the ELF protects the IP rights of their academic collaborators against the pharmaceutical companies, ensuring that the academic researchers can always search for external partners in case that no development deal between them and one of the ELF industry partners could be fixed. This was one of the main concerns when the project started in 2013 [19].

It remains to be seen though if and how academic groups really benefit from these ambitious initiatives, especially when own research interests show little overlap with the essentially commercial interests of the participating pharmaceutical companies.

2.4 The European research infrastructure consortium (ERIC) EU-OPENSCREEN

EU-OPENSCREEN [20] is a community-driven, bottom-up initiative in Europe, which brings together 21 partner sites, i.e. technology platforms and research groups at various universities and research institutions, to form an open-access research infrastructure for chemical biology and early drug discovery. Instead of building an ivory tower, the aim of EU-OPENSCREEN is to establish and operate an infrastructure that facilitates and encourages the engagement with the broader scientific community. In the framework of EU-OPENSCREEN, the partner sites and external researchers collaboratively develop novel tool compounds (or chemical 'probes') that allow researchers to interrogate and study fundamental cellular processes, such as signaling or metabolic pathways.

EU-OPENSCREEN is one of 55 research infrastructures listed on the current ESFRI (European Strategy Forum on Research Infrastructures) Roadmap [21] as an 'ESFRI Landmark Project', demonstrating the relevance for the European scientific community and the European Research Area (ERA). It is jointly funded by the research ministries of eight countries (the Czech Republic, Denmark, Finland, Germany, Latvia, Norway, Poland, Spain) and the European Commission. Since April 2018, it operates a European, not-for-profit organization ('European Research Infrastructure Consortium'), which is based in Berlin, Germany, and is legally independent from any university or research institute. EU-OPENSCREEN, and the European Research Infrastructures in general, promote open science and open innovation [22].

Many technology platforms at universities and research institutes predominantly work with the colleagues at their hosting institution. Larger European initiatives often engage with scientists from Western European countries, where these initiatives are based. Reaching out to, and encourage the active participation of, scientists from regions, which are often underrepresented in chemical biology and early drug discovery research, requires a different approach. Through its distributed network of partner sites across its member countries, EU-OPENSCREEN aims to have a more balanced engagement of local science communities. In each member country, a local partner establishes and coordinates a national network—e.g. CZ-OPENSCREEN in the Czech Republic, PL-OPENSCREEN in Poland, NOR-OPENSCREEN in Norway, Drug Discovery and Chemical Biology Consortium (DDCB) in Finland, ChemBioNet in Germany—to raise awareness about the initiative and to efficiently encourage scientists at the local level to participate.

2.4.1 The research infrastructure

The EU-OPENSCREEN infrastructure provides open-access to compound libraries, assay development and screening facilities, and medicinal chemistry and informatics platforms. It provides training and serves as a platform for industry engagement.

2.4.1.1 Compound collection

The EU-OPENSCREEN compound collection is a diversity library, which has been designed in a collaborative effort of several partner sites. The library is jointly used by affiliated EU-OPENSCREEN partner sites for primary screening against biological targets solicited from external researchers who developed the appropriate assays. During the design of the library, 100,000 commercially available compounds were selected, with an emphasis on chemical stability, absence of reactive compounds, screening-compliant physico-chemical properties, and maximal diversity/coverage of chemical space. Furthermore, EU-OPENSCREEN crowd-sources compounds from external chemists worldwide, in a federated approach through its national chemical biology networks. This collection of academic compounds will, over time, add increasing uniqueness to the EU-OPENSCREEN compound collection. The ambitious goal is to gather up to 40,000 compounds over the next years and to realize the vision of a truly European compound collection. In this context, the EU-OPENSCREEN compound collection will be dynamic and expanding. In analogy to the 'FAIR' (FAIR stands for findability, accessibility, interoperability, and reusability) data principles (described below), structural compound information and quality control data will be available online in an interoperable format (interoperability), unique identifier codes for each compound will be employed (findability), quality control will ensure the identity and purity of the compounds (reproducibility), and their distribution partner sites where they are accessible to external scientists and used in screening projects (accessibility). All compounds of the collections are carefully characterized and annotated for basic physico-chemical (e.g. solubility, light absorbance and fluorescence) and biological properties (e.g. cytotoxicity, antibiotic activity) by 'profiling' them in a standard panel of assays. These bioprofiling data increase the reliability and reproducibility of screening results, and identify compounds with properties that could potentially perturb specific bioassay read-out technologies (e.g. auto-fluorescence, luciferase inhibition, etc.) in order to reduce false-positive results. For chemists who provide compounds to be incorporated in the compound collection, these profiling data are an important incentive, in addition to the bioactivity data from the screening projects.

The jointly used compound collection is stored centrally by the Compound Collection Management Facility (CCMF) in Berlin, Germany, and aliquots are distributed to the affiliated EU-OPENSCREEN partner sites, which are located in the eight EU-OPENSCREEN member countries. The CCMF is responsible for the acquisition, selection, maintenance and storage of the central collection and quality-controls of the compounds. The CCMF provides the screening and bioprofiling sites with copies of the compound collection (including, where necessary, cherry-picking for confirmatory and counter-screening activities).

2.4.1.2 Database

In many cases, primary screening data—even from publicly funded programs—are not openly accessible by the scientific community. While private organizations, contract research organizations (CROs) and many public-private partnerships do not reveal data on a routine basis, EU-OPENSCREEN is committed to maximizing the re-use and impact of generated bioactivity data for the benefit of the wider scientific community. Therefore, EU-OPENSCREEN's ECBD adheres to the FAIR principles [23] and is closely linked to the ChEMBL [24] database, which will raise the discoverability and re-use of EU-OPENSCREEN's data. Via ECBD and ChEMBL, database users will be drawn from across the global biological and chemical science communities, both from academia and industry. Together with other European life sciences-research infrastructures, EU-OPENSCREEN partners also contribute towards the optimization of technological implementation, integration and interoperability of data and tools within the European Open Science Cloud (EOSC) and participate in the Horizon 2020-funded 'EOSC-Life' project (www.eosc-life.eu/). Another initiative, to which the EU-OPENSCREEN partner

Fraunhofer IME actively contributes, is the Innovative Medicines Initiative (IMI) funded 'FAIRplus' project (https://fairplus-project.eu/), which aims to facilitate the application of FAIR principles to data from certain IMI projects and datasets from pharmaceutical companies.

The ECBD is the central database for the integration of screening data from EU-OPENSCREEN projects with advanced search, analysis, and visualization tools. There will be three levels of data management and access: First, bioactivity data generation of compounds in screening projects, implemented at the individual EU-OPENSCREEN screening sites, using assays provided by the external collaboration partners; second, the integration of these screening datasets from partner sites into the ECBD; and, third, public dissemination of the data through established databases like ChEMBL [24] and PubChem [25, 26]. The ECBD is hosted by Petr Bartunek, the coordinator of CZ-OPENSCREEN, and his team at the Institute of Molecular Genetics of the ASCR in Prague, Czech Republic, who have developed the open data resource Probes & Drugs portal [27] as well as other databases such as the Zebrabase [28]. The e-infrastructure CESNET provides cloud-based hosting, backup and security.

An important aspect in the context of integrating complex and diverse screening data, when dealing with datasets from various affiliated, but legally independent sites that jointly use the compound collection, is the implementation of harmonized data standards and data curation. The ECBD adheres to well-established ontologies and identifiers, for example, the BioAssay Ontology (BAO) [29] for the classification and description of assays, which are commonly used by other similar open data repositories, such as ChEMBL or PubChem BioAssay. Only officially accredited partner sites have permission to upload data into the ECBD and uploaded data will be curated both automatically (e.g. file format, column values) as well as manually (e.g. data inspection) by the ECBD team. In case of ambiguities, the ECBD team contacts the data provider to resolve the issue. The ECBD team provides user support and help desk functions. Webinars on data deposition, the use of ECBD for data searching, visualizations and analysis are planned and dedicated workshops will be organized to demonstrate database users all ECBD capabilities and to share best practices in the community.

A grace period of up to 3 years between the completion of the primary screen and data publication in the EU-OPENSCREEN database is provided, during which the bioactivity datasets are not publicly accessible. This grace period allows for follow-up studies, publication in peer-review scientific journals and securing of intellectual property.

Assay development and screening facilities, and medicinal chemistry groups: EU-OPENSCREEN's affiliated screening partner sites implement the EU-OPENSCREEN high-throughput screening (HTS) and High-content screening (HCS) projects by using the EU-OPENSCREEN chemical compound collection, in collaboration with the external assay developer. They have been operational as local groups collaborating with external researchers over the past years, even before the EU-OPENSCREEN ERIC has been established. A recent publication showcases several successful projects, which have been realized by individual partner sites, as an example of the capabilities and expertise within the research infrastructure [20]. The chemistry groups have an excellent, proven track record in medicinal chemistry and hit-to-lead/tool optimization. As part of the collaborations with external researchers, they provide services ranging from the re-synthesis of hit compounds and chemical optimization by synthesis of focused libraries containing structurally similar analogues, elaboration of structure activity relationships (SAR), and NMR and TOF-LC-MS analytics.

2.4.1.3 Training

The EU-OPENSCREEN partner sites have been operational as local screening platforms for many years. During this time, they predominantly work with their colleagues from the hosting institution and university. By working with the same collaborators over a longer time period, both sides could, over the time, increasingly gain practical experience and build a knowledge base, for example, in developing miniaturized, robust assays which are amendable to screening large compound collections. One of the aims of EU-OPENSCREEN is to enable as-yet under-served and under-represented user communities, which, by definition, did not yet have the opportunity to gain practical experience in these areas. Therefore, EU-OPENSCREEN will offer training courses, for example in assay development and other aspects of high-throughput screening. Furthermore, staff exchanges at established partner sites for scientists from prospective sites in countries that are not yet members of EU-OPENSCREEN promote convergence in technical capacities.

2.4.2 Access to the research infrastructure for external researchers

External scientists have open access to a chemical library, assay development and screening facilities, medicinal chemistry and informatics platforms. There are three main groups of researchers who will benefit from EU-OPENSCREEN:

First, molecular and cell biologists, biochemists, microbiologists, plant biologists etc. who develop assays which are amendable to screening and are interested in developing a chemical 'tool' compound for their biological target or pathway of interest to answer a biological question or, in the case of disease-relevant targets, develop new therapeutic approaches to addressing unmet medical needs for patients. These scientists benefit from the open access to the screening capabilities of EU-OPENSCREEN's screening partner sites. They are encouraged to contact and consult the central office of EU-OPENSCREEN, which acts as a single point of contact for external scientists, prior to submitting a project proposal. Depending on the proposal and project requirements, the central office identifies one or more partner sites within the network, which offer the appropriate technology and expertise. The technical feasibility and scientific novelty will be evaluated. After the project proposal has been accepted, the project is initiated in collaboration with a partner site by transferring the assay onto the screening platform. This process often involves further optimization and miniaturization into a 384-well or 1536-well plate format, with the external scientist, who developed the assay, being actively involved in this process at the screening facility. After the screening of the EU-OPENSCREEN compound collection at the EU-OPENSCREEN screening site, data analysis and hit validation, a list with the validated hits will be available to the assay developer. The validated hits will be further optimized either with an EU-OPENSCREEN chemistry site or, if the assay provider already has an established collaboration for the hit-to-lead/tool optimization, with an external chemist.

Second, organic and medicinal chemists and pharmacologists who seek to expose their compounds to a large number of screens, and thereby a wide range of biological targets. They provide their compounds to EU-OPENSCREEN, so that their compounds are 'bio-profiled' and tested as part of the screening collection at the EU-OPENSCREEN partner sites. As chemists often have only limited opportunities to systematically annotate their compounds, their incentive to provide their compounds to EU-OPENSCREEN is the possibility to identify novel biological activities of their compounds. A similar approach to crowd-sourcing academic compounds has been applied over more than a decade within the French Chimiothèque

Nationale [30]. Another, more recent opportunity for chemists to screen their compounds is the CO-ADD (Community for Open Antimicrobial Drug Discovery) [31, 32] initiative, where chemists can test their compounds for antimicrobial activity against ESKAPE pathogens. These initiatives demonstrate that the prospect of receiving bioactivity data is a strong incentive for chemists to donate, and disclose the structure and associated bioactivity data of, their compounds.

Third, database users who use EU-OPENSCREEN's European Chemical Biology Database (ECBD) to access the bioactivity datasets generated during the screening projects. Importantly, the data will also be accessible through other established open data repositories including the ChEMBL database. Assay providers who screen the EU-OPENSCREEN compound library benefit from the ECBD for their own projects by having access to the public bioactivity data from previous projects, and at the same time, they also contribute to worldwide efforts on open science.

2.4.2.1 Access policy and procedure

The democratization of access to state-of-the-art technology platforms, resources and expertise is the key objective of all European research infrastructure. Importantly, as a European open access research infrastructure, a common access policy and procedure is applied across its network of partner sites. EU-OPENSCREEN is accessible to researchers from academia and industry worldwide. The access to EU-OPENSCREEN by external researchers is in line with the 'European Charter for Access to Research Infrastructures—Principles and Guidelines for Access and Related Services' [33] published by the European Commission in 2016. The charter's guidelines describe three access modes, by which access to research infrastructures may be provided—these are excellence-driven, market-driven and wide access. Excellence-driven access is provided to the majority of scientists who developed an assay and collaborate with EU-OPENSCREEN to implement a screening and/or hit optimization project as well as to chemists who provide their compounds to be incorporated in the EU-OPENSCREEN compound collection. Scientists who use the ECBD will be provided wide access to the bioactivity data.

3. Conclusions

In this book chapter, we described various academic collaboration models which aim to accelerate chemical too discovery. These initiatives differ in many aspects, for example in structure (e.g. individual academic research groups, public-private partnerships, research infrastructures; single-site vs. distributed/multinational), operational model (e.g. closed consortia, open-access research infrastructures), user communities, funding model (e.g. institutional funding, third-party funding over a defined funding period, long-term funding by member countries), access and data publication policies. Each of these initiatives complement each other and supports academic chemical biology and drug discovery.

Acknowledgements

The authors would like to thank Ronald Frank (senior advisor of EU-OPENSCREEN, Berlin) and Anna-Lena Gustavsson (Head of the CBCS node at Karolinska Institutet, Stockholm) for ideas and information related to the content of the manuscript.

Conflict of interest

The authors declare no conflict of interest.

Abbreviations

ASCR	Academy of Sciences of the Czech Republic
CESNET	association of universities of the Czech Republic and the Czech Academy of Sciences, operating the national e-infrastructure for science, research and education
ChEMBL	chemical database of bioactive molecules with drug-like properties, maintained by the European Bioinformatics Institute of the European Molecular Biology Laboratory
ECBD	European Chemical Biology Database
ESKAPE	acronym encompassing the names of six bacterial pathogens commonly associated with antimicrobial resistance
NIH	National Institutes of Health
NMR	Nuclear magnetic resonance
TOF-LC-MS	Time-of-flight liquid chromatography mass spectroscopy

Author details

Bahne Stechmann and Wolfgang Fecke*
EU-OPENSCREEN ERIC, Berlin, Germany

*Address all correspondence to: wolfgang.fecke@eu-openscreen.eu

References

[1] Arrowsmith CH, Audia JE, Austin C, et al. The promise and peril of chemical probes. Nature Chemical Biology. 2015;**11**(8):536-541. DOI: 10.1038/nchembio.1867

[2] Renaud JP, Chari A, Ciferri C, et al. Cryo-EM in drug discovery: Achievements, limitations and prospects. Nature Reviews. Drug Discovery. 2018;**17**(7):471-492. DOI: 10.1038/nrd.2018.77

[3] Coussens NP, Sittampalam GS, Guha R, et al. Assay guidance manual: Quantitative biology and pharmacology in preclinical drug discovery. Clinical and Translational Science. 2018;**11**(5):461-470. DOI: 10.1111/cts.12570

[4] Gautam A, Pan X. The changing model of big pharma: Impact of key trends. Drug Discovery Today. 2016;**21**(3):379-384. DOI: 10.1016/j.drudis.2015.10.002

[5] Everett JR. Academic drug discovery: Current status and prospects. Expert Opinion on Drug Discovery. 2015;**10**(9):937-944. DOI: 10.1517/17460441.2015.1059816

[6] Austin CP, Brady LS, Insel TR, Collins FS. NIH molecular libraries initiative. Science. 2004;**306**(5699):1138-1139. DOI: 10.1126/science.1105511

[7] Schreiber SL, Kotz JD, Li M, et al. Advancing biological understanding and therapeutics discovery with small-molecule probes. Cell. 2015;**161**(6):1252-1265. DOI: 10.1016/j.cell.2015.05.02

[8] Hammarström LG, Jensen AJ. Chemical biology consortium Sweden. ACS Chemical Biology. 2013;**8**(12):2605-2606. DOI: 10.1021 /cb400858v

[9] Kallioniemi OP. The SciLifeLab model explained. Sci AAAS [Internet]; 2017. 4. Available from: https://www.sciencemag.org/advertorials/scilifelab-model-explained

[10] Gustavsson AL, Chorell E. Wiley; 2019 (manuscript submitted)

[11] Martinez Molina D et al. Monitoring drug target engagement in cells and tissues using the cellular thermal shift assay. Science. 2013;**341**(6141):84-87. DOI: 10.1126/science.1233606

[12] Jafari R et al. The cellular thermal shift assay for evaluating drug target interactions in cells. Nature Protocols. 2014;**9**(9):2100-2122. DOI: 10.1038/nprot.2014.138

[13] Available from: https://www.thesgc.org/chemical-probes

[14] Arrowsmith CH et al. The promise and peril of chemical probes. Nature Chemical Biology. 2015;**11**(8):536-541. DOI: 10.1038/nchembio.1867

[15] Müller et al. Donated chemical probes for open science. eLife. 2018;**7**:e34311. DOI: 10.7554/eLife.34311

[16] Available from: https://ec.europa.eu/info/funding-tenders/opportunities/portal/screen/opportunities/topic-details/imi2-2019-17-02

[17] Available from: https://www.europeanleadfactory.eu/

[18] Available from: https://www.imi.europa.eu/projects-results/project-factsheets/esculab

[19] Mullard A. European lead factory opens for business. Nature Reviews. Drug Discovery. 2013;**12**(3):173-175. DOI: 10.1038/nrd3956

[20] Brennecke P et al. EU-OPENSCREEN: A novel collaborative approach to facilitate

chemical biology. SLAS Discovery. 2019;**24**:398-413

[21] ESFRI Roadmap 2018—Strategy Report on Research Infrastructures

[22] Principles of data management and sharing at European Research Infrastructures; 2014

[23] Wilkinson MD et al. The FAIR guiding principles for scientific data management and stewardship. Scientific Data. 2016;**3**:160018

[24] Mendez D et al. ChEMBL: Towards direct deposition of bioassay data. Nucleic Acids Research. 2019;**47**:D930-D940

[25] Kim S et al. PubChem 2019 update: Improved access to chemical data. Nucleic Acids Research. 2019;**47**:D1102-D1109

[26] Li Q, Cheng T, Wang Y, Bryant SH. PubChem as a public resource for drug discovery. Drug Discovery Today. 2010;**15**:1052-1057

[27] Skuta C et al. Probes &Drugs portal: An interactive, open data resource for chemical biology. Nature Methods. 2017;**14**:759-760

[28] Oltova J et al. Zebrabase: An intuitive tracking solution for aquatic model organisms. Zebrafish. 2018;**15**:642-647

[29] Vempati UD et al. Formalization, annotation and analysis of diverse drug and probe screening assay datasets using the BioAssay ontology (BAO). PLoS One. 2012;**7**:e49198

[30] Mahuteau-Betzer F. The French National Compound Library: Advances and future prospects. Medical Science. 2015;**31**:417-422

[31] Cooper MA. A community-based approach to new antibiotic discovery. Nature Reviews Drug Discovery. 2015;**14**:587-588

[32] Desselle MR et al. Institutional profile: Community for Open Antimicrobial Drug Discovery - crowdsourcing new antibiotics and antifungals. Future Science OA. 2017;**3**:FSO171

[33] Candido AD. European Charter for Access to Research Infrastructures. Directorate-General for Research and Innovation, European Union. 2016. DOI: 10.2777/524573

Chapter 9

Chemical Biology Toolsets for Drug Discovery and Target Identification

Ammara Riaz, Azhar Rasul, Iqra Sarfraz, Javaria Nawaz, Ayesha Sadiqa, Rabia Zara, Samreen Gul Khan and Zeliha Selamoglu

Abstract

Chemical biology is the scientific discipline that deals with the application of chemical techniques and often small molecules produced through synthetic chemistry, to the manipulation and study of biological systems. Its working framework ranges from simple chemical entities to complex drugs by employing the principles of biological origin. This chapter particularly focuses on the principles and working models of chemical biology to discover new drug leads. Drug discovery is an extensive and multifaceted complex process. Chemical biology uses both natural and synthetic compounds with the best therapeutic potential and verifies them by employing the best possible chemical toolsets. Screening of compounds is done by the use of phenotypic as well as the target-based screening to identify and characterize the potent hits. After the identification of target, it is characterized, and validated by extensive testing. The next step is the validation of hits obtained, and lead compounds are tested in clinical trials before introducing them for commercial application.

Keywords: chemical biology, drug discovery, target identification, target validation, phenotypic screening

1. Introduction to chemical biology and history

Chemical biology flourished as a discipline of science which makes use of several aspects of chemistry to understand biology [1]. Chemical biology includes a wide range of fundamental problems related to the understanding of complex biological processes by the development of synthetic frameworks to generate selective and active lead compounds [2].

The roots of history of chemical biology lie in the emergence of chemistry and biology as separate disciplines. Chemical biology flourished as a separate discipline of science because of newer challenges and questions for the study of chemical methods employed on living bodies. This branch of study is concerned with advanced molecular concepts of biology harnessed to the use of chemical entities. In spite of the newness of this concept, the history of chemical biology extends up to two centuries, considering the foundations of chemistry and biology. Here only a brief account of history of chemical biology is discussed. Joseph Priestley discovered nitrous oxide gas in 1772 and incubated the mice with "airs" (the gases discovered till that time). He used 10 gases including nitrous oxide on experimental mice. His experiments on mice faced a strong mass discontent from Americans who showed a sympathetic behavior towards animal rights. Thus, the first chemical biologist fell a prey to angry mob due to his experiment on mice [3].

Afterwards, another chemist, Humphry Davy, worked (1778–1829) on the newly isolated and unfamiliar gases at that time. Frightened by the previous experiment, Humphry completely omitted the use of mice and decided to carry out the research on himself. It was not a matter of surprise that one of the gases, carbon monoxide, proved fatal for the scientist, but the pleasant effect of nitrous oxide made him name this gas, "the laughing gas." He also investigated the use of this gas in medical surgeries. Samuel Taylor also documented this gas as a pleasure-making gas [4], but the practical use of this gas in medicine was described in 1844 by an American

Figure 1.
History of chemical biology with its eminent events.

dentist, Horace Wells [5]. In 1998, three scientists, namely, Ferid Murad, Robert Furchgott, and Louis Ignarro, won Nobel Prize for the demonstration of significant role of nitric acid in cell signaling [6].

Wöhler is a well-known scientist in the history of chemical biology. He attempted to lay the basis of chemical biology by carrying out his research on vitalism. He prepared urea from inorganic chemicals and rejected the famous "vital force theory" in 1828 [7]. The next important event in the history of chemical biology "cellular imaging" was revolutionized by utilizing the chemical approaches during the nineteenth century. John Hershel invented the cyanotype process which was brought into practice by Anna Atkins to prepare delicate botanical specimens. This noble lady also published her book entitled as *Photographs of British Algae: Cyanotype Impressions* [8].

Ehrlich (1854–1915) is thought to be the pioneer of the earliest forms of chemotherapy and drug therapy. He carried out numerous experiments on aniline based dyes and proposed the idea of "magic bullets." He said that these magic bullets are capable of targeting specific pathogens. He discovered a chemical compound Salvarsan, a drug used against syphilis. This compound is also called as Ehrlich's 606th compound, it was named so because of the successful compound he discovered after 605 failed target compounds. The discovery of this compound paved a way for the discovery of new chemical entities or the new "magic bullets" [6, 9] (**Figure 1**).

Chemical biology flourished as an eminent scientific discipline due to significant contributions of Koehler (pioneer of various chemical screening approaches), Saghatelian (discovery and characterization of lipids and peptides), Wang (use of chemoproteomics in determination of electrophilically lipidated cellular proteins), and Patti and Northen (metabolomics analysis) [1].

2. Chemical biology tools

2.1 Chemical probes

Chemical probes are the small molecules which bind to the specific targeted sites and initiate their cellular activities. These archetypal tools act as highly valued reagents for molecular- and genetic-level biological research. Chemical probes are helpful in the accurate investigation of biological pathways and their associated targets [10].

2.2 Antisense and RNAi technologies

Many tools have been involved in target validation since the 1980s. Target identification and validation are long procedures. They were mainly based on structure-activity relationship. The drug discovery system becomes the most important approach towards the targeted cells [11]. Traditional antisense and RNA interference (RNAi) technologies are the robust tools used in multidimensional phases to discover and validate the potential drug targets. This approach elaborates the potentially selective cleavage of a targeted messenger RNA. This targeting technique enables the researchers to explore the protein-based expression on phenotypes [12].

2.3 Protein degradation strategies

2.3.1 Induced protein degradation

Induced protein degradation is an event-driven approach which depends on drug binding and eliminating the target protein after tagging it. This approach is gaining attention in recent times because of the selective degradation of the target proteins.

Drug discovery based on small molecules focuses on the loss of function of proteins due to the already-occupied binding sites ultimately making the proteins unable to target. In this approach, there is a need of high drug exposure in vivo to avoid target inhibition conditions which may lead to potentially harmful side effects of that drug. Proteolysis-targeting chimeras (PROTACS) use the cellular quality control setup to degrade the selective proteins as their targets. This protein degradation system reduces the quantity of drug to be exposed to the living systems which are to be used for halting the protein functions. These proteins may belong to regulatory proteins, transcription factors, and scaffolding proteins [13, 14].

2.3.2 Chemoproteomics

Chemoproteomics is employed as a chemical tool for target identification. It can be used to investigate the signal transductions. This particular field of study has flourished as a key technology to characterize the action mechanism of chemical probes and drugs which can act as pharmacological modulators, hence validating the cellular targets of several therapeutic drug candidates. Chemoproteomics can be further characterized as affinity- and activity-based chemical proteomics [15]. In some cases when probe development is a difficult task, multiple kinase inhibitors are used for targeting the kinome effectively [16].

3. Drug discovery

Drug discovery is a hectic multistep procedure comprising of highly systematic approaches to identify, and characterize different compounds leading towards the development of hits and validate them extensively via utilization of chemical toolsets to attain the status of a commercial therapeutic drug status. The important steps of drug discovery are mentioned in **Figure 2**.

3.1 Screening

There are two fundamental approaches which can be used for the purpose of drug discovery, namely, phenotypic screening and target-based screening.

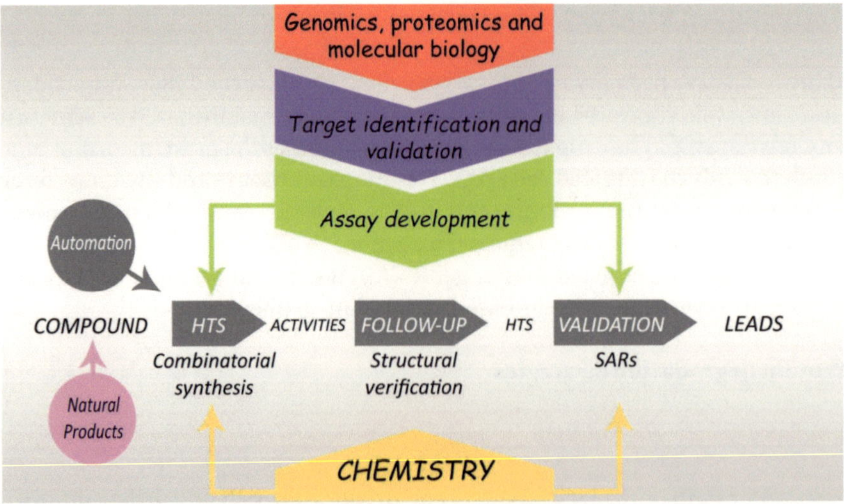

Figure 2.
A diagram representing the summary of key notes regarding drug discovery from natural products.

The first one looks at the effects of phenotype that the compound induces on cell, tissue or whole organism, and the second one evaluates the effects of a compound on a purified target protein.

3.1.1 Phenotypic screening

In the early twentieth century, drug development started with the advancements in pharmacology and synthetic and therapeutic chemistry. In the 1950s and 1960s, enzyme kinetics has provided methods for accurate computation of compound's effectiveness and enzyme competence [17].

Between 1999 and 2008, the US Food and Drug Administration (FDA) approved new drug discovery approaches. During this period, 75 small molecules were discovered and analyzed. Out of these, 28 drugs were discovered through phenotypic selection, and 17 drugs were identified by target dependent selection [18].

"Alemtuzumab" was the first antibody that was been obtained by using hybridoma technology in combination with phenotypic identification. It was previously reported against relapse of multiple sclerosis and chronic lymphocytic leukemia (CLL). The CD44 antigen (cell surface glycoprotein) antagonist, RG7356, was isolated with the help of function F.I.R.S.T™ platform. Therefore, functional assays antibodies were used to check effects on cell signaling, proliferation, and programmed cell death [19].

Large combinatorial antibody libraries are the sources of human monoclonal antibodies, successfully used in medical and phenotypic screening. For example, BI-505 was isolated by using F.I.R.S.T™ platform. Improved versions of antibodies were ultimately used in simulation studies of tumor cell death assay and for selective B-lymphoma cell surface binding. Soon after the isolation of BI-505, its molecular target was identified as ICAM-1, which were found to be involved in apoptosis of B-lymphoma cells. BI-505 has a broad antimyeloma activity [20].

By using phenotypic screening technology, patients can increase their effective antibody response like B-cell repertoire. For example, from a healthcare worker, anti-respiratory syncytial virus (RSV) antibody, D25, was isolated. On the virus coat, D25 neutralizes RSV, and perfusion structure of the F protein was expressed which was not identified by target-based screening [21]. The use of phenotypic screening in various experiments is outlined in **Table 1**.

Disease	Cells	Assay type	Time duration	References
Breast cancer	MCF7-RFP MDA-RFP	Cytochemical and immunohistochemical staining analyses	8–10 days	[22]
Idiopathic pulmonary fibrosis	Alveolar epithelial type II cells	Immunofluorescence staining for in vitro, Western blot, FACs, ELISA, in vitro biochemical kinase assay, migration assay	13 days	[23]
Respiratory papillomatosis	Lung tumor cells	Cell viability assay	48 hours	[24]
Cystic fibrosis	Bronchial epithelial cell	Western blots	18–24 days	[25]
Huntington's disease	PC12	Protease release assay	48 hours	[26]
Familial dysautonomia	Neural crest precursors	RT-PCR assay	48 hours	[27]

Table 1.
Phenotypic screening used in some experiments.

3.1.2 Target-based screening

Target-based screening of natural compounds and synthetic chemicals is being considered as a significant innovation for anticancer drug development [28]. In 2007, Lysine demethylase 5B (KDM5B) and Histone demethylase were recognized, which are liable for the removal of H3K4me2/3 activation marker. Thus, for cancer therapy, KDM5B is regarded as a promising drug target, but the elevated levels of KDM5B were found in many human cancers [29].

The respiratory chain of *Streptococcus agalactiae* consists of two enzymes; type 2-NADH dehydrogenase (NDH-2) and cytochrome *bd* oxygen reductase. *S. agalactiae* is considered as the primary cause of sepsis and meningitis in neonates as well as considerable cause of pneumonia and urinary tract infection [30]. The difference between phenotypic and target-based screening is shown in **Figure 3**.

Some of the target-based screening methods are mentioned as follows.

3.1.2.1 Mass spectrometry-based method

Mass spectrometry is known to be a highly efficient technique for the identification and structural characterization of natural products derived from herbal medicine [31].

Target-based method relies on mass spectrometry to search for active compounds, and this technology can be used for identification, structural characterization, quantitative elemental analysis, tracking of key intermediate compounds in a chemical reaction, analysis of pharmaceuticals and metabolites, and elucidation of unknown structures in drug development. All these achievements can be finally used in various applications like pharmaceutics (drug developments, pharmacokinetics, metabolic pathways), clinical screening, etc. On the basis of MS data information of compounds, the UniFi™ platform has been built for more detailed analysis of structures [32].

3.1.2.2 Liquid chromatography-mass spectrometry (LC-MS)

LC-MS is an analytical technique for separating different complex mixtures into their components using liquid chromatography. These assays check the correct synthesis, purity, various physical and chemical properties like their volatility and active functionalities present in the newly synthesized chemical entities [33]. During drug discovery, LC-MS hyphenated technique is used for seperation and structural characterization of compounds [34].

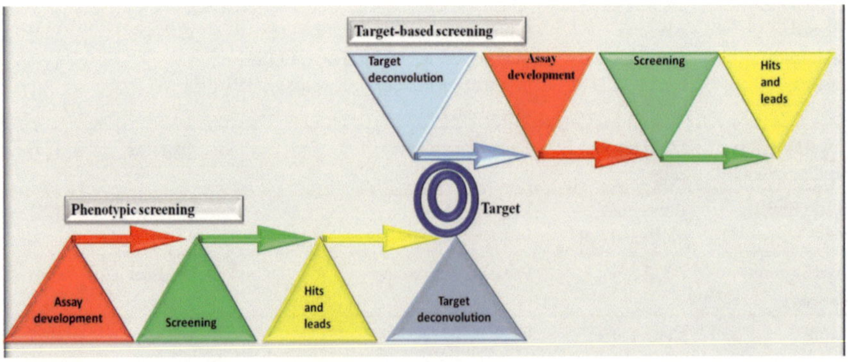

Figure 3.
The action potential of phenotypic as well as target-based screening of compounds to validate the hits and leads from natural and synthetic compounds.

3.1.2.3 Gas chromatography-mass spectrometry (GC-MS)

GC-MS is another hyphenated technique for the identification and structure elucidation of unknown compounds derived from natural products [35]. For example, by using GC-MS technique, comprising a gas chromatograph (GC) coupled to a mass spectrometer (MS), complex components of natural oils mixtures may be separated, identified, and quantified, e.g., oils extracted from Apiaceae family (*Anethum graveolens*, *Carum carvi*, *Cuminum cyminum*, *Coriandrum sativum*, *Pimpinella anisum*, *Daucus carota*, *Apium graveolens*, *Foeniculum vulgare*, and *Ammi visnaga*). As a result of this separation technique, petroselinic acid was the major fatty acid from all other palmitic, palmitoleic, stearic, petroselinic, linoleic, linolinic, and arachidic acids [36].

3.1.2.4 Ultra-performance liquid chromatography-mass spectrometry (UPLC-MS)

Currently, UPLC-MS is one of the most adaptable hyphenated techniques. Proteomics and metabolomics have proved to be useful concepts for understanding the causes of different diseases. This technology aims to seperate and identify proteins and metabolites for cellular signaling pathways and to discover biomarkers for screening and diagnosis as well as determining response to a specific treatment [37]. For example, vancomycin (VCM) is clinically used for the treatment of human intracranial infections. The treatment concentration of vanomycin greatly varies among the patients. UPLC-MS technique was developed and used for the analysis of VCM in human cerebrospinal fluid [38].

3.1.2.5 Nuclear magnetic resonance spectroscopy (NMR)

Among the common techniques of metabolomics, NMR has evolved the most. Unlike mass spectroscopy, NMR is also used for quantitative analysis, but it does not require extra steps for sample preparation [39]. It is commonly used to analyze the 3D structures of biomacromolecules and their interactions. It has been proved a valuable tool for the reliable identification of small molecules that bind to proteins and for hit-to-lead optimization. Mainly, NMR spectroscopy is suitable for the analysis of bulk metabolites [40]. NMR has been used for analyzing the structure of protein, nucleic acid, and small molecule [41]. NMR has been proven to be a useful tool in target-based drug discovery in the step of hit identification and lead optimization [42]. For example, NMR spectroscopy is used to understand the structure of G-quadruplexes, which are noncanonical, four standard nucleic acids with consecutive sequences of guanines [43].

3.1.2.6 Thermal shift or calorimetry-based method

Isothermal titration calorimetry (ITC) is the only technique which is currently available for the direct determination of enthalpy, ΔH, of a ligand binding to a protein [44]. Thermodynamic evaluation might be useful to provide information about specificity, agonist versus antagonist effects of ligands, and other important properties [45]. Fragment-based drug discovery (FBDD) is an approach of particular interest and relevance here. Fragments are molecules smaller than typical drugs, and they generally bind with lower affinity than conventional drug screening hits [46]. Measuring the contributions of enthalpy and entropy to the free energy of binding provides information that can be useful in fragment elaboration and subsequent medicinal chemistry work [47]. ITC is a uniquely powerful tool for characterization of the thermodynamics of test compounds binding to target proteins. Interaction between the compound and protein leads to release or uptake of small amounts of heat, while the mixture is held at a close approximation to

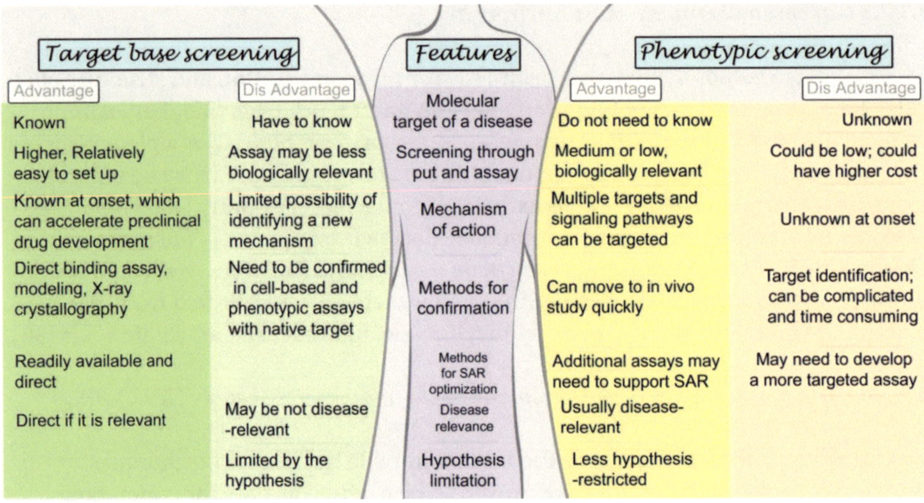

Figure 4.
Comparison between the advantages and disadvantages of target-based and phenotypic screening based upon the different features such as molecular target of disease, its mechanism of action, confirmation methods, SAR optimization methods, and hypothesis limitation.

constant temperature [48]. Thermal shift screening methods has allowed to identify compounds that interact with *Trypanosoma brucei* choline kinase (TBCK) and inhibit TBCK, a validated drug target against African sleeping sickness [49].

3.1.2.7 Affinity-based methods

The methods regarding affinity-based immobilized proteins have vital role in understanding the connections between small molecules and their biological targets [50]. Affinity-based technologies are divided into two groups: (1) direct detection of noncovalent macromolecule-ligand complex and (2) indirect detection of noncovalent macromolecule-ligand complex. The negative aspect of this approach is that it recognizes chemical entities basically based on their binding affinities for a target irrespective of whether or not the biological function of the target is affected. In the late 1980s, matrix-assisted laser desorption ionization (MALDI) and electrospray ionization (ESI) techniques were used to analyze proteins and nucleic acids. Both phenotypic screening and target-based screening are comparable to each other in terms of benefits and drawbacks. This fact has been illustrated in **Figure 4**.

4. Target identification and characterization

Target identification and elucidation of its action mechanism have played vital roles in probing small molecules and drug discovery. Target identification has been based on biological and technologically advanced cell-based assays [51].

4.1 Disease association and target validation

Identification of the molecules and their underlying pathophysiological mechanisms contribute towards the discovery of targets that can be modulated therapeutically [52]. Each drug target is linked to a disease using integrated genome-wide data from a broad range of data sources. The target validation reveals the evidence that associates a target with a disease [53].

4.2 Bioactive small molecules

Bioactive small molecules are preferred as lead structures for the target validation. These small molecules isolated from phenotypic screen play a crucial role in chemical biology [54, 55]. Many genomic, proteomic, and bioinformatic technologies have been developed for validation of the drugs.

4.3 Protein interactions

To identify the selective potent drugs, the first step is to find the protein interference. In signal transductions, protein-protein interactions are involved in the complex cellular networks that govern the different processes [56]. The deregulated transcription factors are involved in playing significant roles in human pathological abnormalities, but the complicated nature of protein-protein networks has made the transcription-targeted therapeutics impractical. Recent technological advancements are the ray of hope regarding the modulation of protein interaction networks [57].

4.4 Cell-based models and target validation

Exosomes are highly adequate for drug carriers as a cell-based model. Due to the association of multiple proteins with cellular membranes, the exosomes are well-known in cell to cell communication, and they are the novel approach for the delivery of potent drugs. Exosome-based drug technique is applied for a variety of disorders such as cancer and various neurodegenerative disorders [58].

5. Target validation

Drug target discovery and validation demand complicated and expensive frameworks which may pose heavy financial load on pharmaceutical industry. Target validation is referred to as the direct involvement of a certain molecular target in pathological conformity; hence, its reversal or inflection may have a therapeutic effect [12].

5.1 Approaches to target validation

The following approaches are used in target validation during the discovery and development of drug.

5.1.1 Antibodies

Firstly access the antibody fitness towards a specific target. Then, standardized procedures are obligatory to ensure the quality of the sample in test procedures; hence, utilizing only a single approach will not work in all situations [59]. Mass spectrometry is used to identify the validation of the antibody. This type of technique confirms the validity for antibodies or their fragments against the targets. The antibody is able to bind to its natural antigen in cell lysates among thousands of other proteins, DNA, RNA, and other cellular components [60].

5.1.2 Cellular thermal shift assay (CETSA)

CETSA is used to assess the capability of a ligand to bind with its targets (cells or tissue samples). The basis of this method lies on the ligand-induced thermodynamic stabilization of target proteins. The compound-treated cell lysates and intact cells were heated to different temperatures, and in the soluble fractions, the target protein was separated from destabilized protein and detected by Western blotting. SPROX is a method of target validation based on identification of ligand-induced stabilization of target proteins. It evaluates the levels of methionine oxidation of target proteins [61].

5.1.3 Drug affinity responsive target stability (DARTS)

DARTS has been used for the identification of the targeted proteins. It is based on ligand binding interaction with proteins forming a complex which changes the structural stability of target protein. There alteration is measured by SDS page/ liquid chromatography. DARTS is also involved in the analysis of the low affinity interactions [61].

6. Hit generation

Hit identification is considered as the significant bottleneck for lead generation success and for new medicines. An example for random hit identification is physical and biochemical testing [62]. The journey of a compound from the hit status to lead status follows a series of steps which have been briefly illustrated in **Figure 5**. The figure describes a note of possible techniques which could be utilized for the selection of lead compounds and proceeding them through lead optimization preclinical and clinical phase trials.

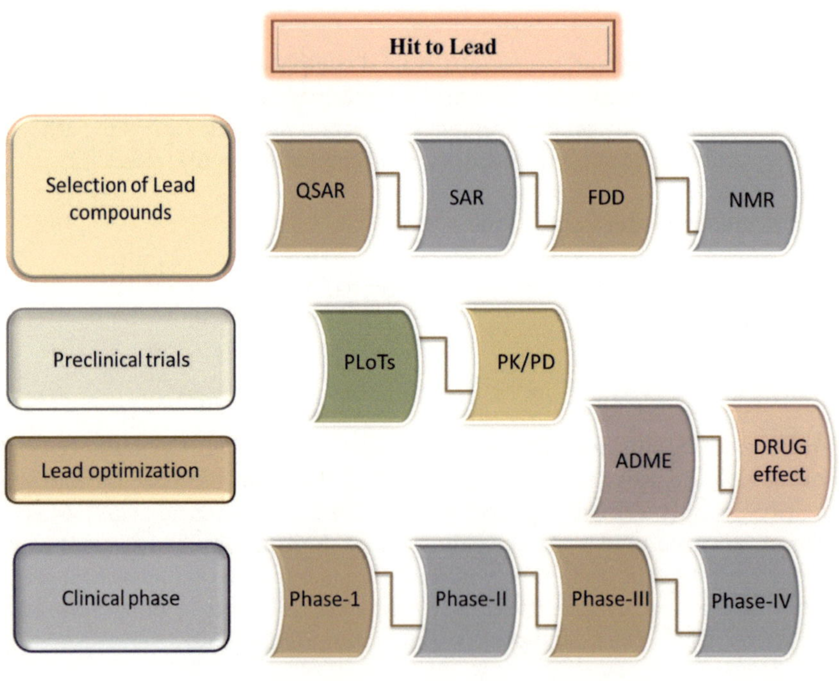

Figure 5.
A diagram elaborating the significant steps of lead optimization proceeding to clinical phase of natural compounds.

7. Development of lead drug

Pharmaceutical companies are facing constant economic pressure to bring efficacy in drug discovery and development process. Lists of compounds obtained after hit optimization are further subjected to refining process in order to find out the lead compounds that can be analyzed for production at commercial scale. During this "hit-to-lead" refining process, many compounds are dropped out due to inadequate absorption, distribution, metabolism, excretion, and toxicity/ADMET characteristics [63].

Refining of hit compounds to lead compound is done through the process of secondary screening. Almost 50% of all drug candidates thin out during optimization and preclinical and clinical trials [64].

There are many approaches available for the discovery and development of drug which might follow different pathways to optimize the compounds into bioavailable drugs. All these pathways must have a common origin; they all begin with a lead compound. It is necessary to go through the phylogeny study of all the compounds because there are some properties like solubility, target affinity, toxicity, ease of synthesis, and bioavailability, all of which are highly dependent on the initial lead selection and the method of identification [65].

7.1 Techniques of lead selection

A rational approach is used to select lead drug candidate after optimization of hit compounds. There are many methods which can be used for screening of compounds. Selection of techniques depends upon the source of hit compounds and types of their solvents as well. The following techniques are useful in selection.

7.1.1 QSAR model development

Quantitative structure-activity relationship model is used to compare chemical structures by using database of prior selected active compounds. Different software like ChemBioOffice Ultra 1.11 is used to generate two-dimensional and three-dimensional structures. The results of QSAR can be validated by using statistical approaches like correlation coefficient and regression coefficient [66].

7.1.2 Visualization of SAR activity

It is called as Bayesian approach. It provides with proficient understanding of shape features, hydrophobic nature, and electrostatic properties of the compounds. All of these features lie under the structure–activity relationship of selected compounds from hits. Structure data analysis of SAR is obtained in 3D form. Other results are obtained in diverse type of interrelated biochemical data, i.e., average of activities and region explored analysis. The results obtained from average activity show a common part in active compounds, and region explored data exhibit the areas of fully explored compounds [67].

7.1.3 Fragment-based drug discovery

It is a powerful method which is used to find out the proportion of ligands with high affinity to target proteins. The compounds which are found to have low ligand binding ability are eliminated, and the compounds with high ligand ability move forward to the precision of compounds. FBDD consists of the techniques such as NMR, SAR, X-ray crystallography, and surface plasmon resonance (SPR).

7.1.3.1 X-ray crystallography

It can ascertain the binding sites and modes of ligand binding to protein [68].

7.1.3.2 Surface plasmon resonance (SPR)

Surface plasmon resonance is known as a nonlabel technology that can identify, screen, and quantify intermolecular interactions in actual time. It is applied to quantify binding affinities. SPR-dependent biosensors work by detecting the ligands and immobilized target molecular interactions and supply appropriate information on kinetics of biomolecular interactions. The output information can be utilized to provide comprehensive functional data on binding actions such as specificity, kinetics, concentration, and affinity [69]. Scientific literature study revealed Biacore tools as mainly used SPR technology at commercial levels [70].

7.2 Preclinical trials

In the last 2 years, different methodologies based on high-throughput screening and their combinations with chemistry have been developed in order to manufacture versatile compounds by limiting the resources. Among these methodologies, several other in vitro and in silico supplementary approaches have also come forward for the identification and potential evaluation of these compounds as lead candidate validation. Those compounds which are selected as "hits" during this screening procedure are further analyzed and subjected to in vivo toxicity and efficacy profiling. During preclinical stage of drug development, simple formulation approaches are favored. Combinatorial chemistry and high-throughput approaches have been appraised in several publications [71].

PLOTs are preclinical lead optimization technologies that should be rapid enough to edge with high-throughput discovery screenings without causing further delay and should be predictive and cost-effective. PLOT platform usually comprised of in vitro systems, small and acquiescent to mechanization, and that is why it is easy to achieve the mandatory throughput with minimum use of compound use [72].

7.2.1 Tools of preclinical drug development

Selection of methodology and tools for selection of preclinical drug candidates is a rigorous process. Sequential approach of preclinical to clinical is practiced to sort out the long list of target selected compounds. This streamline strategy provides with deeper understanding of action of the drug prior to its progress to the next steps [73].

7.2.2 Pharmacokinetics and pharmacodynamics (PK/PD) during preclinical drug evaluation

Pharmacodynamics involves the study of effect of drug in dose- and time-dependent manner. Pharmacokinetics is the study of absorption, metabolization, distribution, and excretion of a drug over time. PK/PD is a program at early phase of lead drug development which acts as a bridge between drug discovery and preclinical drug development. This stage set aims for further development activities, and information obtained at this stage act as a key to subsequent steps.

It is necessary because of the following reasons:

a. It provides potency-based intrinsic activity of the compound rather than dose.

b. It characterizes the compounds on the basis of dose concentration and effect relationship.

c. It allows the investigation of tolerance phenomenon of compounds on the basis of physiological parameters [74].

7.3 Lead optimization

Optimization of a drug is a multifaceted process. It usually involves various types of screening methods which tend to find out the metabolism and pharmacokinetic properties of selected compounds or drugs [75].

7.3.1 ADME

This is the final stage of preclinical trials; after this the optimized drug is further processed towards the clinical trial. Absorption, distribution, metabolism, and excretion screening is performed at this stage. The primary goal of ADME is to develop a competitive drug with adequate safety avoiding PK failure in clinical phase.

7.3.2 ADME properties

Ideal properties of a drug in ADME testing involve the good oral bioavailability, blood clearance and volume of convenient dosing, and low potential of drug-drug interaction. All of these properties are assessed at early stage of drug discovery [76].

7.3.3 DRUGeff

Drug effect is a parameter which determines the concentration of a drug which do not cause any harm at the site of action. In other words at this stage, toxicity of a drug is tested to find out the minimum safe dosage potency. In vitro DRUGeff testing of all compounds show interaction with the target treatment, until a small portion of dose gets to select according to biophase levels. Concentration of treatment dose maximization per unit of biophase acts as a key objective for lead optimization. The drugs qualifying this test enter into the clinical phase [77].

7.4 Clinical phase of drug discovery

The final step of drug discovery and development is referred to as the clinical trial. At this stage, the data regarding safety and efficacy of the new drug must be proven by application to humans directly in different phases. After the successful trials, research data is sent to the FDA for approval for commercial manufacturing and marketing (**Figure 6**) [78].

7.4.1 Clinical phase I

The first phase of clinical trial normally takes several weeks to some months. At this stage application of optimized drug is tested on a small group of volunteers.

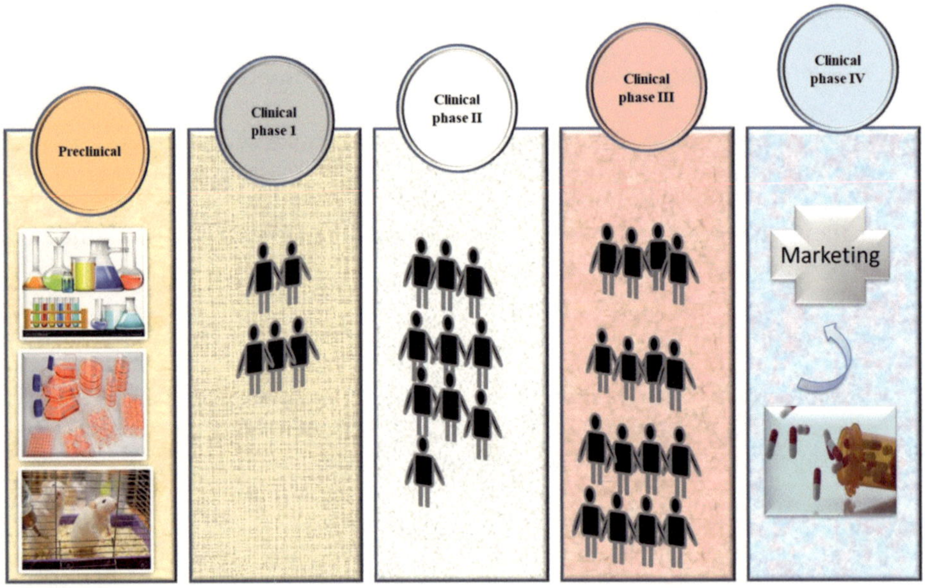

Figure 6.
The journey of potential leads from preclinical to clinical trials.

They may or may not get paid for their participation in drug trial studies. This mini trial is useful in determining the absorption and side effect of drug in relation to its dose concentration [17].

7.4.2 Clinical phase II

The second phase of clinical trial may last up to 2 years. It is a totally randomized study which involves the application of drug on a relatively large group of patients. This trial study is divided into two groups of patients, one receiving experimental drug and the other receiving placebo. Sometimes it may be named as a blind application trial. This type of random application of drug allows investigators and pharmaceutics to prove the success and safety of drug to the FDA with comparative information [79].

7.4.3 Clinical phase III

It is a large-scale testing of drugs on hundreds of patients. This third stage testing provides with a more thorough understanding and effectiveness of useful drugs to the FDA and pharmaceutical companies. The pharmaceutical company can request for the approval for commercial synthesis of drug after phase III is completed [80].

7.4.4 Clinical phase IV

After the approval of a drug for commercial consumption, clinical phase IV trials are used as post marketing surveillance trials. This trial system is based upon the various objectives at commercial levels, i.e., the comparison of newly approved and already-available drugs in market, to evaluate the chronic effects on patients' quality of life and to estimate the economical comparison of newly approved and already-present drugs as well as the traditional system of medication [81].

8. Conclusion and future perspectives

Chemical biology is an emerging field of science which particularly focuses on the research in biological systems by employing the chemicals and related chemoinformatic tools. This field of study is working well in combination with medicinal and combinatorial chemistry to seek the cure of incurable and life-threatening human pathologies. This chapter illustrated the significant techniques and chemical setups which can be employed to testify the chemical as well as biological aspects of natural and synthetic compounds before introducing them as therapeutic drugs in the field of medicine. There is an ultimate need of the hour to seek for the newer and better drugs which are safer, cheaper, and more effective than the already existing therapeutics. This field of study is flourishing at a very fast pace, and it is anticipated that it will provide better treatment options and strategies in future for the medical practitioners to use the best among the rest drugs discovered.

Conflict of interest

Authors have no conflict of interest.

Author details

Ammara Riaz[1], Azhar Rasul[1*], Iqra Sarfraz[1], Javaria Nawaz[1], Ayesha Sadiqa[1], Rabia Zara[1], Samreen Gul Khan[2] and Zeliha Selamoglu[3]

1 Department of Zoology, Faculty of Life Sciences, Government College University Faisalabad, Faisalabad, Pakistan

2 Department of Chemistry, Faculty of Physical Sciences, Government College University Faisalabad, Faisalabad, Pakistan

3 Department of Medical Biology, Faculty of Medicine, Nigde Ömer Halisdemir University, Nigde, Turkey

*Address all correspondence to: drazharrasul@gmail.com

References

[1] Saghatelian A, Nomura DK, Weerapana E. Editorial overview: Omics: The maturation of chemical biology. Current Opinion in Chemical Biology. 2016;**30**:v-vi

[2] Ostler EL. Chemical biology is. Chemistry Central Journal. 2007;**1**:5

[3] Priestley J. Experiments and Observations on Different Kinds of Air: And Other Branches of Natural Philosophy, Connected with the Subject. In Three Volumes: Being the Former Six Volumes Abridged and Methodized, with Many Additions. United Kingdom: Thomas Pearson; 1790

[4] Hoover SR. Coleridge, Humphry Davy, and some early experiences with a consciousness-altering drug. Bulletin of Research in the Humanities. 1978;**81**:9-27

[5] Wright AJ. Davy comes to America: Woodhouse, Barton, and the nitrous oxide crossing. Journal of Clinical Anesthesia. 1995;**7**:347-355

[6] Morrison KL, Weiss GA. The origins of chemical biology. Nature Chemical Biology. 2006;**2**:3-6

[7] Wöhler F. Poggendorff's. Annals of Physical Chemistry. 1828;**12**:253

[8] Atkins A, Chuang J, Schaaf L. Anna Atkins: Photographs of British Algæ. Germany: Gerhard Steidl Druckerei und Verlag; 2020

[9] Miescher F. Die histochemischen und physiologischen Arbeiten von Friedrich Miescher. Leipzig, Germany: Vogel; 1897

[10] Blagg J, Workman P. Choose and use your chemical probe wisely to explore cancer biology. Cancer Cell. 2017;**32**:9-25

[11] Duarte Y, Marquez-Miranda V, Miossec MJ, Gonzalez-Nilo F. Integration of target discovery, drug discovery and drug delivery: A review on computational strategies. Wiley Interdisciplinary Reviews. Nanomedicine and Nanobiotechnology. 2019;**11**:e1554

[12] Lavery KS, King TH. Antisense and RNAi: Powerful tools in drug target discovery and validation. Current Opinion in Drug Discovery & Development. 2003;**6**:561-569

[13] Lai AC, Crews CM. Induced protein degradation: An emerging drug discovery paradigm. Nature Reviews. Drug Discovery. 2017;**16**:101-114

[14] Toure M, Crews CM. Small-molecule PROTACS: New approaches to protein degradation. Angewandte Chemie. 2016;**55**:1966-1973

[15] Drewes G, Knapp S. Chemoproteomics and chemical probes for target discovery. Trends in Biotechnology. 2018;**36**:1275-1286

[16] Yao Z, Petschnigg J, Ketteler R, Stagljar I. Application guide for omics approaches to cell signaling. Nature Chemical Biology. 2015;**11**:387-397

[17] King TA, Stewart HL, Mortensen KT, North AJP, Sore HF, Spring DR. Cycloaddition strategies for the synthesis of diverse heterocyclic spirocycles for fragment-based drug discovery. European Journal of Organic Chemistry. 2019;**2019**:5219-5229

[18] Swinney DC, Anthony J. How were new medicines discovered? Nature Reviews. Drug Discovery. 2011;**10**:507-519

[19] Menke-van der Houven van Oordt CW, Gomez-Roca C, van Herpen C, Coveler AL, Mahalingam D, Verheul HM, et al. First-in-human phase I clinical trial of RG7356, an anti-CD44 humanized antibody, in patients with advanced, CD44-expressing

solid tumors. Oncotarget. 2016;7:80046-80058

[20] Fransson J, Tornberg UC, Borrebaeck CA, Carlsson R, Frendeus B. Rapid induction of apoptosis in B-cell lymphoma by functionally isolated human antibodies. International Journal of Cancer. 2006;**119**:349-358

[21] McLellan JS, Chen M, Leung S, Graepel KW, Du X, Yang Y, et al. Structure of RSV fusion glycoprotein trimer bound to a perfusion-specific neutralizing antibody. Science. 2013;**340**:1113-1117

[22] Khan GN, Kim EJ, Shin TS, Lee SH. Heterogeneous cell types in single-cell-derived clones of MCF7 and MDA-MB-231 cells. Anticancer Research. 2017;**37**:2343-2354

[23] Fujino N, Kubo H, Maciewicz RA. Phenotypic screening identifies Axl kinase as a negative regulator of an alveolar epithelial cell phenotype. Laboratory Investigation. 2017;**97**:1047-1062

[24] Yuan H, Myers S, Wang J, Zhou D, Woo JA, Kallakury B, et al. Use of reprogrammed cells to identify therapy for respiratory papillomatosis. The New England Journal of Medicine. 2012;**367**:1220-1227

[25] Fulcher ML, Gabriel SE, Olsen JC, Tatreau JR, Gentzsch M, Livanos E, et al. Novel human bronchial epithelial cell lines for cystic fibrosis research. The American Journal of Physiology-Lung Cellular and Molecular Physiology. 2009;**296**:L82-L91

[26] Titus SA, Southall N, Marugan J, Austin CP, Zheng W. High-throughput multiplexed quantitation of protein aggregation and cytotoxicity in a Huntington's disease model. Current Chemical Genomics. 2012;**6**:79-86

[27] Lee G, Ramirez CN, Kim H, Zeltner N, Liu B, Radu C, et al. Large-scale screening using familial dysautonomia induced pluripotent stem cells identifies compounds that rescue IKBKAP expression. Nature Biotechnology. 2012;**30**:1244-1248

[28] Yamori T. Chemical evaluation by cancer cell line panel and its role in molecular target-based anticancer drug screening. Cancer Chemotherapy. 2004;**31**:485-490

[29] Rotili D, Mai A. Targeting histone demethylases: A new avenue for the fight against cancer. Genes & Cancer. 2011;**2**:663-679

[30] Yamamoto Y, Pargade V, Lamberet G, Gaudu P, Thomas F, Texereau J, et al. The group B streptococcus NADH oxidase Nox-2 is involved in fatty acid biosynthesis during aerobic growth and contributes to virulence. Molecular Microbiology. 2006;**62**:772-785

[31] Henke MT, Kelleher NL. Modern mass spectrometry for synthetic biology and structure-based discovery of natural products. Natural Product Reports. 2016;**33**:942-950

[32] Wang X, Zhang A, Han Y, Wang P, Sun H, Song G, et al. Urine metabolomics analysis for biomarker discovery and detection of jaundice syndrome in patients with liver disease. Molecular & Cellular Proteomics. 2012;**11**:370-380

[33] Lee MS, Kerns EH. LC/MS applications in drug development. Mass Spectrometry Reviews. 1999;**18**:187-279

[34] Yu S, Li S, Yang H, Lee F, Wu JT, Qian MG. A novel liquid chromatography/tandem mass spectrometry based depletion method for measuring red blood cell partitioning of pharmaceutical compounds in drug discovery. Rapid Communications in Mass Spectrometry. 2005;**19**:250-254

[35] Spanik I, Machynakova A. Recent applications of gas chromatography

with high-resolution mass spectrometry. Journal of Separation Science. 2018;**41**:163-179

[36] Nguyen T, Aparicio M, Saleh MA. Accurate mass GC/LC-quadrupole time of flight mass spectrometry analysis of fatty acids and triacylglycerols of spicy fruits from the Apiaceae Family. Molecules. 2015;**20**:21421-21432

[37] Zhao YY, Lin RC. UPLC-MS(E) application in disease biomarker discovery: The discoveries in proteomics to metabolomics. Chemico-Biological Interactions. 2014;**215**:7-16

[38] Mei S, Wang J, Zhu L, Chen R, Li X, Chen K, et al. A UPLC-MS/MS method for analysis of vancomycin in human cerebrospinal fluid and comparison with the chemiluminescence immunoassay. Biomedical Chromatography. 2017;**31**:e3939

[39] Emwas AH. The strengths and weaknesses of NMR spectroscopy and mass spectrometry with particular focus on metabolomics research. Methods in Molecular Biology. 2015;**1277**:161-193

[40] Harner MJ, Frank AO, Fesik SW. Fragment-based drug discovery using NMR spectroscopy. Journal of Biomolecular NMR. 2013;**56**:65-75

[41] Billeter M, Wagner G, Wuthrich K. Solution NMR structure determination of proteins revisited. Journal of Biomolecular NMR. 2008;**42**:155-158

[42] Pellecchia M, Sem DS, Wuthrich K. NMR in drug discovery. Nature Reviews. Drug Discovery. 2002;**1**:211-219

[43] Lin C, Dickerhoff J, Yang D. NMR studies of G-Quadruplex structures and G-Quadruplex-interactive compounds. Methods in Molecular Biology. 2019;**2035**:157-176

[44] Wiseman T, Williston S, Brandts JF, Lin LN. Rapid measurement of binding constants and heats of binding using a new titration calorimeter. Analytical Biochemistry. 1989;**179**:131-137

[45] Chaires JB. Calorimetry and thermodynamics in drug design. Annual Review of Biophysics. 2008;**37**:135-151

[46] Hopkins AL, Groom CR, Alex A. Ligand efficiency: A useful metric for lead selection. Drug Discovery Today. 2004;**9**:430-431

[47] Ward WH, Holdgate GA. Isothermal titration calorimetry in drug discovery. Progress in Medicinal Chemistry. 2001;**38**:309-376

[48] Recht MI, De Bruyker D, Bell AG, Wolkin MV, Peeters E, Anderson GB, et al. Enthalpy array analysis of enzymatic and binding reactions. Analytical Biochemistry. 2008;**377**:33-39

[49] Major LL, Denton H, Smith TK. Coupled enzyme activity and thermal shift screening of the Maybridge rule of 3 fragment library against Trypanosoma brucei choline kinase; a genetically validated drug target. In: El-Shemy HA, editor. Drug Discovery. Rijeka (HR): IntechOpen; 2013. pp. 413-431

[50] Temporini C, Brusotti G, Pochetti G, Massolini G, Calleri E. Affinity-based separation methods for the study of biological interactions: The case of peroxisome proliferator-activated receptors in drug discovery. Methods. 2018;**146**:12-25

[51] Schenone M, Dancik V, Wagner BK, Clemons PA. Target identification and mechanism of action in chemical biology and drug discovery. Nature Chemical Biology. 2013;**9**:232-240

[52] Floris M, Olla S, Schlessinger D, Cucca F. Genetic-driven druggable target identification and validation. Trends in Genetics. 2018;**34**:558-570

[53] Koscielny G, An P, Carvalho-Silva D, Cham JA, Fumis L, Gasparyan R, et al.

Open targets: A platform for therapeutic targetidentificationandvalidation.Nucleic Acids Research. 2017;**45**:D985-D994

[54] Velagapudi SP, Gallo SM, Disney MD. Sequence-based design of bioactive small molecules that target precursor microRNAs. Nature Chemical Biology. 2014;**10**:291-297

[55] Jung HJ, Kwon HJ. Target deconvolution of bioactive small molecules: The heart of chemical biology and drug discovery. Archives of Pharmacal Research. 2015;**38**:1627-1641

[56] Rimbault C, Maruthi K, Breillat C, Genuer C, Crespillo S, Puente-Munoz V, et al. Engineering selective competitors for the discrimination of highly conserved protein-protein interaction modules. Nature Communications. 2019;**10**:4521

[57] Mapp AK, Pricer R, Sturlis S. Targeting transcription is no longer a quixotic quest. Nature Chemical Biology. 2015;**11**:891

[58] Batrakova EV, Kim MS. Using exosomes, naturally-equipped nanocarriers, for drug delivery. Journal of Controlled Release: Official Journal of the Controlled Release Society. 2015;**219**:396-405

[59] Taussig MJ, Fonseca C, Trimmer JS. Antibody validation: A view from the mountains. New Biotechnology. 2018;**45**:1-8

[60] Persson H, Preger C, Marcon E, Lengqvist J, Graslund S. Antibody validation by immunoprecipitation followed by mass spectrometry analysis. Methods in Molecular Biology. 2017;**1575**:175-187

[61] Chang J, Kim Y, Kwon HJ. Advances in identification and validation of protein targets of natural products without chemical modification. Natural Product Reports. 2016;**33**:719-730

[62] Bleicher KH, Bohm HJ, Muller K, Alanine AI. Hit and lead generation: Beyond high-throughput screening. Nature Reviews. Drug Discovery. 2003;**2**:369-378

[63] Nyunoya H, Lusty CJ. The carB gene of Escherichia coli: A duplicated gene coding for the large subunit of carbamoyl-phosphate synthetase. Proceedings of the National Academy of Sciences of the United States of America. 1983;**80**:4629-4633

[64] Lofas S. Optimizing the hit-to-lead process using SPR analysis. Assay and Drug Development Technologies. 2004;**2**:407-415

[65] Fejzo J, Lepre CA, Peng JW, Bemis GW, Ajay, Murcko MA, et al. The SHAPES strategy: An NMR-based approach for lead generation in drug discovery. Chemistry & Biology. 1999;**6**:755-769

[66] Floresta G, Rescifina A, Marrazzo A, Dichiara M, Pistara V, Pittala V, et al. Hyphenated 3D-QSAR statistical model-scaffold hopping analysis for the identification of potentially potent and selective sigma-2 receptor ligands. European Journal of Medicinal Chemistry. 2017;**139**:884-891

[67] Alam S, Khan F. QSAR and docking studies on xanthone derivatives for anticancer activity targeting DNA topoisomerase II alpha. Drug Design, Development and Therapy. 2014;**8**:183-195

[68] Erlanson DA, Davis BJ, Jahnke W. Fragment-based drug discovery: Advancing fragments in the absence of crystal structures. Cell Chemical Biology. 2019;**26**:9-15

[69] Liu C, Yang Y, Wu Y. Recent advances in exosomal protein detection via liquid biopsy biosensors for cancer screening, diagnosis, and prognosis. The AAPS Journal. 2018;**20**:41

[70] Kukanskis K, Elkind J, Melendez J, Murphy T, Miller G, Garner H. Detection of DNA hybridization using the TISPR-1 surface plasmon resonance biosensor. Analytical Biochemistry. 1999;**274**:7-17

[71] Ramstrom O, Lehn JM. Drug discovery by dynamic combinatorial libraries. Nature Reviews. Drug Discovery. 2002;**1**:26-36

[72] Atterwill CK, Wing MG. In vitro preclinical lead optimisation technologies (PLOTs) in pharmaceutical development. Toxicology Letters. 2002;**127**:143-151

[73] Boger E, Friden M. Physiologically based pharmacokinetic/pharmacodynamic modeling accurately predicts the better bronchodilatory effect of inhaled versus oral salbutamol dosage forms. Journal of Aerosol Medicine and Pulmonary Drug Delivery. 2019;**32**:1-12

[74] Ekblom M, Hammarlund-Udenaes M, Paalzow L. Modeling of tolerance development and rebound effect during different intravenous administrations of morphine to rats. Journal of Pharmacology and Experimental Therapeutics. 1993;**266**:244-252

[75] Cheng KC, Korfmacher WA, White RE, Njoroge FG. Lead optimization in discovery drug metabolism and pharmacokinetics/case study: The hepatitis C virus (HCV) protease inhibitor SCH 503034. Perspectives in Medicinal Chemistry. 2007;**1**:1-9

[76] Balani SK, Miwa GT, Gan LS, Wu JT, Lee FW. Strategy of utilizing in vitro and in vivo ADME tools for lead optimization and drug candidate selection. Current Topics in Medicinal Chemistry. 2005;**5**:1033-1038

[77] Braggio S, Montanari D, Rossi T, Ratti E. Drug efficiency: A new concept to guide lead optimization programs towards the selection of better clinical candidates. Expert Opinion on Drug Discovery. 2010;**5**:609-618

[78] Swann PG, Shapiro MA. Regulatory considerations for development of bioanalytical assays for biotechnology products. Bioanalysis. 2011;**3**:597-603

[79] Sartori SB, Singewald N. Novel pharmacological targets in drug development for the treatment of anxiety and anxiety-related disorders. Pharmacology & Therapeutics. 2019;**204**:107402

[80] Regan D, Garcia K, Thamm D. Clinical, pathological, and ethical considerations for the conduct of clinical trials in dogs with naturally occurring cancer: A comparative approach to accelerate translational drug development. ILAR Journal. 2018;**59**:99-110

[81] Stephenson N, Shane E, Chase J, Rowland J, Ries D, Justice N, et al. Survey of machine learning techniques in drug discovery. Current Drug Metabolism. 2019;**20**:185-193

Chapter 10

Artificial Intelligence-Based Drug Design and Discovery

Yu-Chen Lo, Gui Ren, Hiroshi Honda and Kara L. Davis

Abstract

The drug discovery process from hit-to-lead has been a challenging task that requires simultaneously optimizing numerous factors from maximizing compound activity, efficacy to minimizing toxicity and adverse reactions. Recently, the advance of artificial intelligence technique enables drugs to be efficiently purposed *in silico* prior to chemical synthesis and experimental evaluation. In this chapter, we present fundamental concepts of artificial intelligence and their application in drug design and discovery. The emphasis will be on machine learning and deep learning, which demonstrated extensive utility in many branches of computer-aided drug discovery including de novo drug design, QSAR (Quantitative Structure–Activity Relationship) analysis, drug repurposing and chemical space visualization. We will demonstrate how artificial intelligence techniques can be leveraged for developing chemoinformatics pipelines and presented with real-world case studies and practical applications in drug design and discovery. Finally, we will discuss limitations and future direction to guide this rapidly evolving field.

Keywords: artificial intelligence, chemoinformatics, data mining, drug discovery

1. Introduction

The path of drug discovery from small molecule ligands to drugs that can be utilized clinically has been a long and arduous process. Starting with a hit compound, the drugs need to be evaluated through multiple *in vitro* and cell-based assays to improve the mechanism of actions followed by mouse models to demonstrate appropriate *in vivo* and transport properties. Mechanistically, the drugs not only need to exert enough binding affinity to the disease targets, but also necessitate proper transport through multiple physiological barriers to enable access to these targets. Other problems like chemical toxicity, often induced by off-targets interactions with unintended proteins as well as pharmacogenetic, where genetic variation influences drug responses all need to be considered in drug design. Therefore, these multifaceted problems in drug discovery often posed significant challenges for drug designers. Recently, the rise of artificial intelligence approach saw potential solutions to these challenges. A sub-umbrella of artificial intelligence called machine-learning has taken a central stage in many R&D sectors of pharmaceutical companies that allows drugs to be developed more efficiently and at the same time mitigate the cost associated with the required experiments [1]. Given some observations of chemical data, machine learning can be used to construct a predictor by learning compound properties from extracted features of compound structures and interactions. Because this approach does not require a mechanistic understanding of how drugs behave, many compound properties like binding affinity and other transport and toxicity problems can be accurately forecasted in this way before they are synthesized [2]. Furthermore, by simultaneously tackling the Pharmacokinetics/Pharmacodynamics (PK/PD) problems using artificial intelligence, we can expect that the effort and time required to bring a drug from bench to bedside can be substantially reduced. In this regard, the artificial intelligence approach has now become an essential tool to facilitate the drug discovery process.

2. Chemoinformatic for drug discovery

2.1 Chemical formats

To facilitate the discussion on artificial intelligence and machine learning in drug discovery and design, it is necessary to understand the type of format and data presentation commonly used for chemical compounds in chemoinformatics. Chemoinformatics is a broad field that studying the application of computers in storing, processing and analyzing chemical data. The field already has more than 30 years of development with focuses on subjects such as chemical representation, chemical descriptors analysis, library design, QSAR analysis and computer-aided drug design [3]. Along with these developments, several popular chemical data formats for data processing has been proposed. Intuitively, the chemical compound is best represented by graphs, also known as "chemical graph" or "molecular graph" where nodes represent atoms and edges represent bonds. The molecular graph is useful for distinguishing different structural isomers but does not contain 3D conformation of the molecules. To store 2D or 3D coordinates of compounds, chemical file formats such as Structure Data Format (SDF), MDL (Molfile), and Protein Data Bank (PDB) formats can be used. In contrast to the PDB file that simply store structural data, the SDF format provides additional advantages of recording descriptors and other chemical properties thus offers better functionality for cheminformatics analysis. Due to the limited memory capacity for handling large compound database, several chemical line notations have also been introduced. One such format is the simplified molecular-input line-entry system (SMILES) format pioneered by Weininger et al [4]. Other linear notations include Wiswesser line notation (WLN), ROSDAL, and SYBYL Line Notation (SLN). Instead of recording compound coordinates directly, the SMILES format store compound structure using simpler ASCII codes. While memory-efficient, there is no unique strings for representing chemical compound particularly for large and structurally complex molecules. To address this, canonical SMILES was proposed that applied the Morgan algorithm for consistent labeling and ordering of chemical structures [5]. Another limitation is the loss of coordinate information and necessitate structural generation programs like PRODRG to predict native molecular geometry [6]. Recently, the need to exchange chemical data over the world wide web (WWW) also saw the development of chemical markup language (CML) similar to the XML format. Despite the development of multiple chemical file formats, many commercial and open source packages have allowed convenient file format conversion using Obabel and RDKit softwares [7, 8].

2.2 Chemical representations

The ability to represent chemical compounds by machine-learning features that fully captured wide ranges of chemical and physical properties of the target molecule has been an active area of research in chemoinformatics and chemical biology

[9, 10]. These chemical features, also known as chemical descriptors, provide the ability to extract essential characteristic of the compound and offer the possibility of developing predictor that can classify novel structures with similar properties. Broadly speaking, the chemical descriptors can be classified as 0D, 1D, 2D, 3D, and 4D [11]. 0D and 1D descriptors like molecular mass, atom number counts can be easily extracted from the molecular formula but does not provide much discriminatory power for compound classification. In practice, 2D and 3D chemical descriptors are the most commonly used molecular features for cheminformatics analysis [12]. Since chemical compound can be viewed as different arrangements of atoms and chemical bond, 2D descriptors can be generated from the molecular graph based on different connectivity of the molecules. Notable 2D descriptors include Weiner index, Balaban index, Randic index and others [1]. Beyond 2D descriptors, 3D descriptors leverage information from molecular surfaces, volumes, and shapes to provide a higher level of chemical representation. The dependency of ligand conformations also prompts the development of 4D descriptors, which accounts for different conformations of the molecules generated over a trajectory from the molecular dynamics simulation [13]. However, the requirement of correct 3D conformation makes 3D and 4D descriptors limited in several aspects. Another type of high dimensional descriptors is molecular interaction field (MIF) developed by Goodford and colleagues [14]. The MIF aims to capture the molecular environment of the ligand based on several properties by placing probes in a rectangular grid surround the target compound. At each grid point, hypothetical probes corresponding to different types of energetic interactions (hydrophobic, electrostatic) were evaluated. The comparison of MIF of compounds enables the identification of critical functional groups for kinase drug-target interactions and drug design [15]. Furthermore, correlating these field values to compound activity enable comparative molecular field analysis (CoMFA), an extended form of 3D-QSAR [16]. Altman's group at Stanford University took a different approach by inspecting ligand environment using amino acid microenvironment. This Feature-based approach lead to direct applications in pocket similarity comparison for identifying novel microtubule binding activity of several anti-estrogenic compounds as well as kinase off-target binding activity [17, 18]. Chemical descriptors can likewise be generated based on the biological phenotypes. For example, drug-induced cell cycle profile changes of compound have been recently utilized to identify DNA-targeting properties of several microtubule destabilizing agents [19].

Besides chemical descriptors, the chemical fingerprint is another important chemical representation where the compounds are represented by a binary vector indicating the presence or absence of chemical features [20]. Common 2D chemical fingerprints include path-based fingerprint which detected all possible linear paths consisting of bonds and atoms of a structure given certain bond lengths. For a given pattern, several bits in a bit string is set. While path-based fingerprints like ECFP (Extended Connectivity Fingerprint) have a higher specificity, the potential limitation is "bit collision" where the number of possible patterns exceeds the bit capacity resulting in multiple patterns mapped to the same set of bits. Another type of fingerprint is substructure fingerprints. In the substructure fingerprint like (Molecular ACCess System) MACCS keys, the substructures are predefined and each bit in a bit string is set for specific chemical patterns. Although bit collision is less of an issue, the requirement to encompass all fragment space within a bit string often demands a larger memory size. Recently, the proposal of circular fingerprints represents the state-of-the-art in chemical fingerprint development [21]. In the circular fingerprint, each layer's feature is constructed by applying a fixed hash function to the concatenated features of the neighborhood in the previous layer and the results from the hashed function were mapped to bit string representing specific substructures. A modified version of the circular fingerprint, known as graph convolution fingerprint, has recently been proposed where the hashed function is replaced by a differential neural network and a local filter is applied to each atom and neighborhoods similar to that of a convolution neural network. Many of the mentioned fingerprints has been implemented by several open source chemoinformatics package such as Chemoinformatics Development Kit (CDK) and RDKit and saw wide applications in compound database search and other computer-aided drug discovery tasks [22].

3. Artificial intelligence in drug discovery

The rise of artificial intelligence and, in particular, machine learning and deep learning has given rise to a tsunami of applications in drug discovery and design [23, 24]. Here, we provide an overview of machine learning concepts and techniques commonly applied for chemoinformatics analysis. In a nutshell, machine learning aims to build predictive models based on several features derived from the chemical data, many of which are measured experimentally, such as lipophilicity, water solubility while others are purely theoretical, such as chemical descriptors and molecular fields derived from the chemical graph or 3D structure data. With chemical features on one hand, on the other hand of the equation is the properties that the model intended to learn, which can take on categorical or continuous values and usually pertaining to compound activity in question. Given every pair of features and labels, the model can be trained by identifying an optimal set of parameters that minimizes certain objective functions. Following the training phase, the best model can then be applied to predict the properties of new compounds (**Figure 1**).

Although machine learning has just recently gained in popularity, its application in chemistry is not new. The pioneering work of Alexander Crum-Brown and Thomas Fraser in elucidating the effects of different alkaloids on muscle paralysis results in the proposal of the first general equation for a structure–activity relationship, which intended to bridge biological activity as a function of chemical structure [25]. Early QSAR models such as Hansch analysis were mostly linear or quadratic model of physicochemical parameters that required extensive experimental measurement. This model was succeeded by the Free-Wilson model, which considers the parameters generated from the chemical structure and is more closely resemble the QSAR model in use today. Machine learning techniques in cheminformatics analysis can be broadly classified as supervised learning, unsupervised learning, and reinforcement learning. However, new learning algorithms through a combination of these approaches are continuing being developed. Many of these approaches have already found wide application in QSAR/QSPR prediction, de novo drug design, drug repurposing, and retrosynthetic planning [26–28].

3.1 Supervised learning

3.1.1 Linear regression analysis

Supervised learning has a long history of development in QSAR analysis [29]. The supervised learning task can include classification, to determine whether a compound class belong to a certain class label, or regression, to predict the bioactivity of a compound over a continuous range of values. A well-known supervised learning approach is the linear regression model, and often the first-line method for exploratory data analysis among statistician. The goal of linear regression is to find

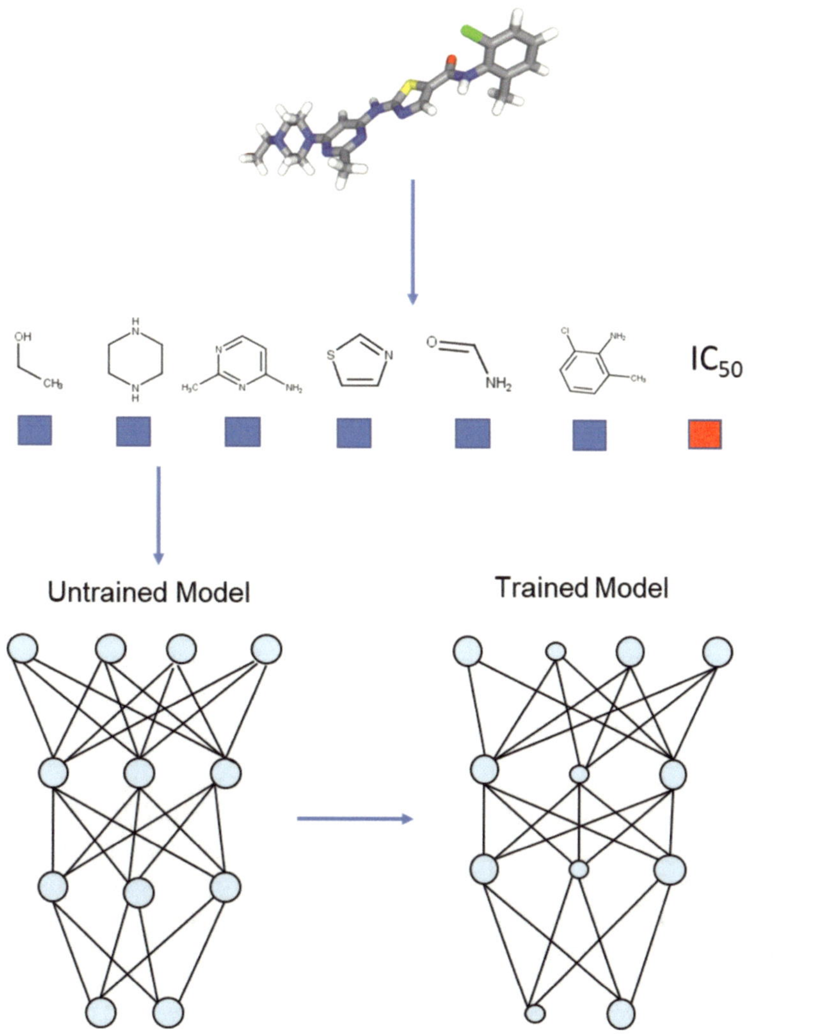

Figure 1.
Chemoinformatics prediction using artificial intelligence. Starting with a compound, the chemical feature is extracted from the compound 2D graph. The chemical features then serve as input for the machine learning model and trained based on the compound activity. The trained model with fitted parameters can then be used to predict activity of new compounds.

a linear function such that a fitted line that minimizes the distance to the outcome variables. When the logistic function is applied to the linear model, the model can also be applicable for binary classification. A direct extension of linear regression is polynomial regression that model relationships between independent and independent variable as high-degree polynomial of the same or different combination of chemical features. In the case of model underfitting, polynomial regression provides a useful alternative for feature augmentation for the linear model. Both linear and polynomial regression formed the basis of classical Hansch and Free-Wilson analysis [30]. Interestingly, today's situation is completely reversed. With the rapid explosion of chemical descriptors and fingerprints available at chemoinformatician's disposal, twin curse of dimensionality and collinearity has now become a significant issue.

Several approaches have been developed to tackle high dimensional data. One potential solution is to exhaustively explore all the possible combination of features to identify the best subset of predictors. However, this approach is inevitably computationally infeasible for large feature space. To solve this, heuristic approach like forward and backward feature selection were developed where each feature was added to the predictors in a stepwise manner and only features that contribute greatest to the fit are kept [31]. An alternative approach for feature selection is dimensional reduction where a smaller set of uncorrelated features can be created as a combination of a larger set of correlated variables. One commonly used dimensional reduction technique is principal component analysis (PCA) that identifies new variables with the largest variances in the dataset [32]. Recently, variable shrinkage method like regularization and evolutionary algorithm has allowed feature selection during the model fitting phase. In the model regularization step, a penalty term is introduced to the objective function to control model complexity. The lasso regularization is one such approach that used an L1 penalty term to constraint objective function along the parameter axis, thus enable effective elimination of redundant features [33]. The evolutionary algorithm is another feature selection approach that encodes features as genes and through successive combination, the algorithm identifies the best set of features measured by a fitness score. Recently, elastic net combines penalties of the lasso and ridge regression and shows promise in variable selection when the number of predictors (p) is much bigger than the number of observations (n) [34]. Although linear regression analysis formed the backbone of early QSAR analysis, the simple linear assumption of feature vector space is a major limitation for modeling more complex system.

3.1.2 Artificial neural network and deep learning

The requirement to parameterize the QSAR model in a non-linear way saw the widespread application of artificial neural network (ANN) in the chemoinformatic analysis. The ANN, first developed by Bernard Widrow of Stanford University in the 1950s, is inspired by the architecture of a human brain, which consisting of multiple layers of interconnecting nodes analogous to biological neurons. The early neural network model is called "perceptron" that consists of a single layer of inputs and a single layer of output neurons connected by different weights and activation functions [35]. However, it was soon recognized that the one-layer perceptron cannot correctly solve the XOR logical relationship [36]. This limitation prompts the development of multi-layer perceptron, where additional hidden layers were introduced into the model and the weights were estimated using the backpropagation algorithm [37]. As a direct extension of ANN, several deep learning techniques like deep neural network (DNN) has been introduced to process high dimensional data as well as unstructured data for machine vision and natural language processing (NLP). In multiple studies, DNN outperformed several classical machine learning methods in predicting biological activity, solubility, ADMET properties and compound toxicity [38, 39].

To handle high-dimensional data, several feature extraction and dimension reduction mechanisms has been integrated into diverse deep learning frameworks (**Figure 2**). In particular, the convolution neural network is a popular deep learning framework for imaging analysis [40]. A convolution neural network consists of convolution layers, max-pooling layers, and fully connected multilayer perceptron. The purpose of the convolution and max-pooling layer is to extracted local recurring patterns from the image data to fit the input dimension of the fully connected layers. This utility has recently been extended for protein structure analysis in the 3D-CNN approach where protein structures are treated as 3D images [41]. Other deep learning approaches include autoencoder and embedding representation. Autoencoder (AE) is a data-driven approach to obtain a latent presentation of high dimensional data using a smaller set of hidden neurons [42, 43]. An autoencoder consists of encoder and decoder. In the encoding step, the input signal is forward propagated to smaller and smaller sets of hidden layers thus effective map the data to low dimensional space. The training is achieved so that the hidden layers can propagate back to a larger set of output nodes to recover the original signal. A specific form of AE called variational AE (VAE) has recently been applied to de-novo drug design application where latent space was first constructed from the ZINC database from which novel compounds can be recovered by sampling such subspace [44]. In the context of NLP, word embedding such as word2vec implementation is a dimensional reduction technique to learn word presentation that preserves the similarity between data in low-dimension. This formulation has been extended to identify chemical representation in the analogous mol2vec program [45]. The requirement to model sequential data also prompted the development of recurrent neural networks (RNN). The RNN is a variant of artificial neural network where the output from the previous state is used as input for the current state. Therefore, this formulation has a classical analogy to the hidden Markov model (HMM), a type of belief network. RNN has been applied for de novo molecule design by "memorizing" from SMILES string in sequential order and generated novel SMILES by sampling from the underlying probability distribution [46]. By tuning the sampling parameters, it is found that RNN can oftentimes generated valid SMILES string not found in the original training set.

3.1.3 Instance-based learning

In contrast to parametrized learning that required extensive efforts in model tuning and parameter estimation, instance-based learning, also known as memory-based learning, is a different type of machine learning strategy that generates hypothesis from the training data directly [47]. Therefore, the model complexity

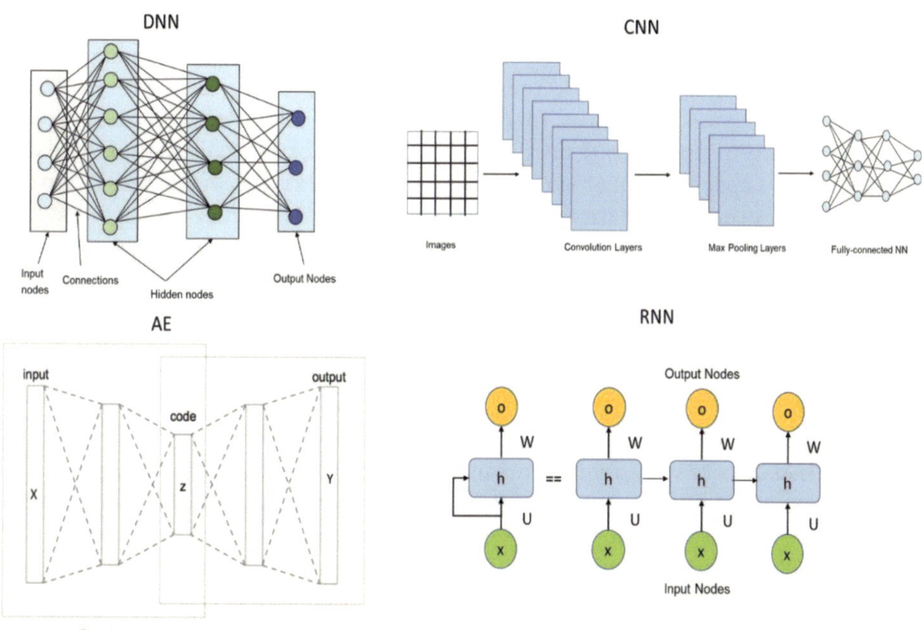

Figure 2.
Deep learning architectures for drug discovery. Four common types of deep learning network for supervised and supervised learning including deep neural network (DNN), convolutional neural network (CNN), autoencoder (AE) and recurrent neural network (RNN).

is highly dependent on the size and quality of the dataset. Notable instance-based learning method includes the k-Nearest Neighbor (kNN) prediction, commonly known as "guilt-by-association" or "like-predicts-like". In the kNN algorithm, a majority voting rule is applied to predict the properties of a given data, based on the k nearest neighbor within certain metric distance [48]. Using this approach, the properties of the data can be inferred from the dominant properties shared among its nearest neighbors. In the field cheminformatics, chemical similarity principle is a direct application of kNN where the similarity between chemical structures can be used to infer similar biological activity [49]. For analyzing large compound set, chemical similarity networks, or chemical space networks, can be used to identify chemical subtypes and estimate chemical diversity [50, 51]. Furthermore, the similarity concept is commonly applied in computational chemical database search to identify similar compounds from a lead series [52]. A major limitation of kNN is the correct determination of the number of nearest neighbors since that too high or low of such parameter can lead to either high false positive and false negative rates.

In the case of binary classification, such as compound activity discrimination, support vector machine (SVM) is a popular non-parametrized machine learning model [53]. For given binary data labels, SVM intended to find a hyperplane such that it has the largest distance (margin) to the nearest training data point of two classes. Furthermore, kernel trick allows mapping data points to high dimensional feature space that are linearly inseparable. For multilabel classification problems, other instance-learning models such as radial basis neural network (RBNN), decision trees and Bayesian learning are generally applicable [54]. In RBNN, several radial basis functions, which often depict as bell shape regions over the feature space, are used to approximate the distribution of the data set. Other approaches like decision tree, such as the Classification And Regression Tree (CART) algorithm, can also be applied for multi-variable classification and regression and has been used to differentiate active estrogen compound from inactives [55]. In the decision tree model, the algorithm provides explanations for the observed pattern by identifying predictors that maximize the homogeneity of the dataset through successive binary partitions (splits). The Bayesian classifier is yet another powerful supervised learning approach that predicts future events based on past observations known as prior. In essence, Bayes' theorem allows the incorporation of prior probability distributions to generate posterior probabilities. In the case of multi-variable classification, a special form of Bayesian learner known as the naïve Bayes learner greatly simplify the computational complexity with independence assumption between features. PASS Online is an example of a Bayesian approach to predict over 4000 kinds of biological activity, including pharmacological effects, mechanisms of action, toxic and adverse effects [56]. In another study, DRABAL, a novel multiple label classification method that incorporates structure learning of a Bayesian network, was developed for processing more than 1.4 million interactions of over 400,000 compounds and analyze the existing relationships between five large HTS assays from the PubChem BioAssay Database [57].

While instance-based learning encompasses a diverse set of methodology and present unique advantages in constantly adapting to new data, this approach is nevertheless limited by the memory storage requirement and, as the dataset grows, data navigation becomes increasingly inefficient. To address this, data pre-segmentation technique such as KD tree is a common approach for instance reduction and memory complexity improvement [58]. In another aspect, the ability to assemble different classifiers into a meta-classifier that will potentially have superior generalization performance than individual classifier also led to the development of ensemble learning. The ensemble learning algorithm can include models that combine multiple types of classifier or sub-sample data from a single model. A notable example of ensemble learning is the random forest algorithm, which combines multiple decision trees and makes predictions via a majority voting rule for compound activity classification and QSAR modeling [59].

3.2 Unsupervised learning

Given a compound dataset, unsupervised learning can include tasks such as detecting subpopulation to determine the number of chemotypes to estimate chemical diversity and chemical space visualization. Putting in a broader perspective, the purpose of unsupervised learning is to understand the underlying pattern of the datasets. Another important problem stem from unsupervised learning is the ability to define appropriate metrics that can be used to quantify the similarity of data distributed over feature space. These metrics can be useful for chemometrics application including measuring the similarity between pairs of compounds.

3.2.1 Clustering

For unsupervised clustering, one popular approach is K-means clustering [60]. K-means clustering aims to partition the dataset into K-centroid. This is achieved by constantly minimizing the within-cluster distances and updating new centroids until the location of the K-centroids converges. K-means clustering has the advantage of operating at linear time but does not guarantee convergence to a global minimum. Another limitation is the requirement of a pre-determined number of clusters, which may not correspond to the optimal clusters for the data. To identify the optimal k values, one solution is called the "elbow method", which determine a k value with the largest change in the sum of distances as the k value increases. One study applied K-means clustering to estimate the diversity of compounds that inhibit cytochrome 3A4 activity [61]. Besides K-mean clustering, conventional clustering like hierarchical clustering is also commonly used. Hierarchical clustering can include agglomerative clustering, which merges smaller data objects to form larger clusters or divisive clustering, which generate smaller clusters by splitting from a large cluster. The hierarchical clustering has been demonstrated for their ability to classify large compound and enrich ICE inhibitors from specific clusters as well as for virtual screening application [62, 63].

Although hierarchical clustering is suitable for initial exploratory analysis, it is limited by several shortcomings such as high space and time complexity and lack of robustness to noise. Supervised clustering using artificial networks include the self-organization map (SOM), also known as Kohonen network [64]. The purpose of SOM is to transform the input signal into a two-dimensional map (topological map) where input features that are similar to each other are mapped to similar regions of the map. The learning algorithm is achieved by competitive learning through a discriminant function that determines the closest (winning) neuron. During each training iteration, the winning neuron has its weight updated such that it moves closer to the corresponding input vector until the position of each neuron converges. The advantages of SOM are the ability to directly visualize the high-dimensional data on low dimensional grid. Furthermore, the neural network makes SOM more robust to the noisy data and reduces the time complexity to the linear range. SOMs cover such diverse fields of drug discovery as screening library design, scaffold-hopping, and repurposing [65].

Recently, manifold learning has gained tremendous traction due to the ability to perform dimensional reduction while preserving inter-point distances in lower dimension space for large-scale data visualization. Manifold learning algorithm includes ISOMAP, which build a sparse graph for high dimensional data and identify the shortest distance that best preserves the original distance matrix in low dimensional space [66]. While ISOMAP requires very few parameters, the approach is nevertheless computational expensive due to an expensive dense matrix eigen-reduction process. More efficient approaches such as Locally Linear Embedding (LLE) has been proposed for QSAR analysis [67]. LLE assumes that the high dimensional structure can be approximated by a linear structure that preserves the local relationship with neighbors. A related approach is t-distributed stochastic neighbor embedding (tSNE), which relies on the pair-wise probability distribution of data points to preserve local distance [68].

3.2.2 Similarity

The ability to measure data similarity is as important as the ability to discern the number of categories from a dataset. One approach for measuring data similarity is by determining the distance of two data points in the high-dimensional feature space. Intuitively, the similarity between two data points is inversely related to the measured distance between them. Commonly used distance metrics include Euclidean distance, Manhattan distance, Chebyshev distance [60]. All of these metrics is a specialized form of Minkowski distance, a generalized distance metrics defined in the norm space. Other important similarity measures such as the cosine similarity and Pearson's correlation coefficient, are commonly used to measure gene expression data or word embedding vector, when the magnitude of the vector is not essential. For binary features, metrics that measured shared bits between vectors can be used. For example, Tanimoto index, also known as the Jaccard coefficient, is one of the most commonly used metrics to measuring the similarity between two fingerprints in many cheminformatics applications. Tanimoto index has been extended to measure the similarity of 3D molecular volume and pharmacophore, such as those generated from the ligand structural alignment [69]. A generalized form of similarity metric is the kernel such as RBF or Gaussian kernel, which is a function that maps a pair of input vectors to high dimensional space and is an effective approach to tackle non-linearly separable case for discriminating analysis. The selection of an optimal similarity metrics can be achieved by clustering analysis, including comparing the clustering result and assess the quality of the clusters by different similarity measures.

3.3 Reinforcement learning

Reinforcement Learning came into the spotlight from the famous chess competition between professional chess player and AlphaGo that demonstrated the ability of AI to outcompete human intelligence [70]. Differ from supervised and unsupervised learning, the reinforcement learning focused on optimization of rewards and the output is dependent on the sequence of input. A basic reinforcement learning is modeled based on the Markov decision process and consists of a set of environment and agent state, a set of actions and transitional probability between states. At each time step, the agent interacts with the environment with a chosen action and a given reward. Several learning strategies have been developed to guide the action in each state. The most well-known algorithm is called the Q-learning algorithm [71]. The Q-learning predicts an expected reward of an action in a given state and as the agent interacts with the environment, the Q value function becomes progressively better at approximate the value of an action in a given state. Another approach for guiding the action for reinforcement learning is called policy learning, which aims to create a map that suggests the best action for a given state. The policy can be constructed using a deep neural network. Recently, deep Q-network (DQN) has been constructed that approximate the Q value-functions using a deep neural network [72]. One recent example of using deep reinforcement learning in de novo design is demonstrated by the ReLeaSE (Reinforcement Learning for Structural Evolution), which integrates both predictive and generative model for targeted library design based on SMILES string. The generative model is used to generate chemically feasible compound while the predictive model is then used to forecast the desired properties. The ReLeaSE method can be used to design chemical libraries with a bias toward structural complexity or toward compounds with a specific range of physical properties as well as inhibitory activity against Janus protein kinase 2 [73].

4. Conclusion

The path of drug discovery from small molecule ligand to drug that can be utilized clinically is a long and arduous process. The fundamental concept of artificial intelligence and the application in drug design and discovery presented will facilitate this process. In particular, the machine learning and deep learning, which demonstrated great utility in many branches of computer-aided drug discovery like de novo drug design, QSAR analysis, chemical space visualization.

In this chapter, we presented the fundamental concept of artificial intelligence and their application in drug design and discovery. We first focused on chemoinformatics, a broad field that studying the application of computers in storing, processing, and analyzing chemical data. This field already has more than 30 years of development with focuses on subjects ranging from chemical representation, chemical descriptors analysis, library design, QSAR analysis, and retrosynthetic planning. We then discussed how artificial intelligence techniques can be leveraged for developing more effective chemoinformatics pipelines and presented with real-world case studies. From the algorithmic aspects, we mentioned three major class of machine learning algorithms including supervised learning, unsupervised learning, and reinforcement learning, each with their own strength and weakness as well as cover different areas of chemoinformatic applications.

As AI techniques gradually become indispensable tools for drug designer to solve their day-to-day problems, an emerging trend is to learn how to flexibly integrate these algorithms in the computational pipelines suitable for the problem at hand. For example, the process can start with an unsupervised learning to discerning the number of chemotypes followed by a supervised learning approach to predict multi-target activities. Furthermore, with the increasing computational power, deep learning network with increasing number layers and complexity will be also developed. Another potential development is the marriage between chemical big data and AI to mine the chemical "universe" for drug screening applications. The potential extensibility of AI in drug discovery and design is virtually boundless and awaits drug designer to further explore this exciting field.

Author details

Yu-Chen Lo[1,3*], Gui Ren[2], Hiroshi Honda[2] and Kara L. Davis[3]

1 Bioengineering, Stanford University, Stanford, CA, USA

2 Bioengineering, Northwestern Polytechnic University, Fremont, CA, USA

3 Pediatrics, Bass Center for Childhood Cancer, Stanford School of Medicine, Stanford, CA, USA

*Address all correspondence to: bennylo@stanford.edu

References

[1] Lo YC, Rensi SE, Torng W, Altman RB. Machine learning in chemoinformatics and drug discovery. Drug Discovery Today. 2018;**23**(8):1538-1546. DOI: 10.1016/j.drudis.2018.05.010

[2] Idakwo G, Luttrell J, Chen M, Hong H, Zhou Z, Gong P, et al. A review on machine learning methods for in silico toxicity prediction. Journal of Environmental Science and Health. Part C, Environmental Carcinogenesis & Ecotoxicology Reviews. 2018;**36**(4):169-191. DOI: 10.1080/10590501.2018.1537118

[3] Gasteiger J. Chemoinformatics: A new field with a long tradition. Analytical and Bioanalytical Chemistry. 2006;**384**(1):57-64. DOI: 10.1007/s00216-005-0065-y

[4] Weininger D. SMILES, a chemical language and information system. 1. Introduction to methodology and encoding rules. Journal of Chemical Information and Computer Sciences. 1988;**28**(1):31-36. DOI: 10.1021/ci00057a005

[5] O'Boyle NM. Towards a universal SMILES representation—A standard method to generate canonical SMILES based on the InChI. Journal of Cheminformatics. 2012;**4**(1):22. DOI: 10.1186/1758-2946-4-22

[6] Schuttelkopf AW, van Aalten DM. PRODRG: A tool for high-throughput crystallography of protein-ligand complexes. Acta Crystallographica. Section D, Biological Crystallography. 2004;**60**(Pt 8):1355-1363. DOI: 10.1107/S0907444904011679

[7] O'Boyle NM, Banck M, James CA, Morley C, Vandermeersch T, Hutchison GR. Open babel: An open chemical toolbox. Journal of Cheminformatics. 2011;**3**:33. DOI: 10.1186/1758-2946-3-33

[8] Lovric M, Molero JM, Kern R. PySpark and RDKit: Moving towards big data in cheminformatics. Molecular Informatics. 2019;**38**(6):e1800082. DOI: 10.1002/minf.201800082

[9] Gupta A, Kumar V, Aparoy P. Role of topological, electronic, geometrical, constitutional and quantum chemical based descriptors in QSAR: mPGES-1 as a case study. Current Topics in Medicinal Chemistry. 2018;**18**(13):1075-1090. DOI: 10.2174/156802661866618071919164149

[10] Haggarty SJ, Clemons PA, Wong JC, Schreiber SL. Mapping chemical space using molecular descriptors and chemical genetics: Deacetylase inhibitors. Combinatorial Chemistry & High Throughput Screening. 2004;**7**(7):669-676

[11] Sykora VJ, Leahy DE. Chemical descriptors library (CDL): A generic, open source software library for chemical informatics. Journal of Chemical Information and Modeling. 2008;**48**(10):1931-1942. DOI: 10.1021/ci800135h

[12] Nettles JH, Jenkins JL, Bender A, Deng Z, Davies JW, Glick M. Bridging chemical and biological space: "Target fishing" using 2D and 3D molecular descriptors. Journal of Medicinal Chemistry. 2006;**49**(23):6802-6810. DOI: 10.1021/jm060902w

[13] Pan D, Tseng Y, Hopfinger AJ. Quantitative structure-based design: Formalism and application of receptor-dependent RD-4D-QSAR analysis to a set of glucose analogue inhibitors of glycogen phosphorylase. Journal of Chemical Information and Computer Sciences. 2003;**43**(5):1591-1607. DOI: 10.1021/ci0340714

[14] Goodford PJ. A computational procedure for determining energetically favorable binding sites on biologically important macromolecules. Journal of Medicinal Chemistry. 1985;**28**(7):849-857. DOI: 10.1021/jm00145a002

[15] Naumann T, Matter H. Structural classification of protein kinases using 3D molecular interaction field analysis of their ligand binding sites: Target family landscapes. Journal of Medicinal Chemistry. 2002;**45**(12):2366-2378. DOI: 10.1021/jm011002c

[16] Cramer RD, Patterson DE, Bunce JD. Comparative molecular field analysis (CoMFA). 1. Effect of shape on binding of steroids to carrier proteins. Journal of the American Chemical Society. 1988;**110**(18):5959-5967. DOI: 10.1021/ja00226a005

[17] Lo YC, Liu T, Morrissey KM, Kakiuchi-Kiyota S, Johnson AR, Broccatelli F, et al. Computational analysis of kinase inhibitor selectivity using structural knowledge. Bioinformatics. 2019;**35**(2):235-242. DOI: 10.1093/bioinformatics/bty582

[18] Lo YC, Cormier O, Liu T, Nettles KW, Katzenellenbogen JA, Stearns T, et al. Pocket similarity identifies selective estrogen receptor modulators as microtubule modulators at the taxane site. Nature Communications. 2019;**10**(1):1033. DOI: 10.1038/s41467-019-08965-w

[19] Lo YC, Senese S, France B, Gholkar AA, Damoiseaux R, Torres JZ. Computational cell cycle profiling of cancer cells for prioritizing FDA-approved drugs with repurposing potential. Scientific Reports. 2017;**7**(1):11261. DOI: 10.1038/s41598-017-11508-2

[20] Durant JL, Leland BA, Henry DR, Nourse JG. Reoptimization of MDL keys for use in drug discovery. Journal of Chemical Information and Computer Sciences. 2002;**42**(6):1273-1280

[21] Kearnes S, McCloskey K, Berndl M, Pande V, Riley P. Molecular graph convolutions: Moving beyond fingerprints. Journal of Computer-Aided Molecular Design. 2016;**30**(8):595-608. DOI: 10.1007/s10822-016-9938-8

[22] Willighagen EL, Mayfield JW, Alvarsson J, Berg A, Carlsson L, Jeliazkova N, et al. The chemistry development kit (CDK) v2.0: Atom typing, depiction, molecular formulas, and substructure searching. Journal of Cheminformatics. 2017;**9**(1):33. DOI: 10.1186/s13321-017-0220-4

[23] Mater AC, Coote ML. Deep learning in chemistry. Journal of Chemical Information and Modeling. 2019;**59**(6):2545-2559. DOI: 10.1021/acs.jcim.9b00266

[24] Hessler G, Baringhaus KH. Artificial intelligence in drug design. Molecules. 2018;**23**(10):E2520. DOI: 10.3390/molecules23102520

[25] Klebe G. Drug Design. New York: Springer; 2013

[26] Jordan AM. Artificial intelligence in drug design-the storm before the calm? ACS Medicinal Chemistry Letters. 2018;**9**(12):1150-1152. DOI: 10.1021/acsmedchemlett.8b00500

[27] Jing Y, Bian Y, Hu Z, Wang L, Xie XQ. Deep learning for drug design: An artificial intelligence paradigm for drug discovery in the big data era. The AAPS Journal. 2018;**20**(3):58. DOI: 10.1208/s12248-018-0210-0

[28] Roy K. In Silico Drug Design. Waltham, MA: Elsevier; 2019. p. 886

[29] Gasteiger J. Handbook of Chemoinformatics: From Data to Knowledge. Weinheim: Wiley-VCH; 2003

[30] Cherkasov A, Muratov EN, Fourches D, Varnek A, Baskin II, Cronin M, et al. QSAR modeling: Where have you been? Where are you going to? Journal of Medicinal Chemistry. 2014;**57**(12):4977-5010. DOI: 10.1021/jm4004285

[31] Hastie T, Tibshirani R, Friedman JH. The Elements of Statistical Learning: Data Mining, Inference, and Prediction. 2nd ed. New York, NY: Springer; 2009. xxii. p. 745

[32] Akella LB, DeCaprio D. Cheminformatics approaches to analyze diversity in compound screening libraries. Current Opinion in Chemical Biology. 2010;**14**(3):325-330. DOI: 10.1016/j.cbpa.2010.03.017

[33] Varnek A, Baskin I. Machine learning methods for property prediction in chemoinformatics: Quo Vadis? Journal of Chemical Information and Modeling. 2012;**52**(6):1413-1437. DOI: 10.1021/ci200409x

[34] Zou H, Hastie T. Regularization and variable selection via the elastic net. Journal of the Royal Statistical Society, Series B: Statistical Methodology. 2005;**67**(2):301-320. DOI: 10.1111/j.1467-9868.2005.00503.x

[35] Widrow B, Lehr MA. 30 Years of adaptive neural networks: Perceptron, Madaline, and backpropagation. Proceedings of the IEEE. 1990;**78**(9):1415-1442. DOI: 10.1109/5.58323

[36] Minsky M, Papert S. Perceptrons; an Introduction to Computational Geometry. Cambridge, Mass: MIT Press; 1969. p. 258

[37] Rumelhart DE, Hinton GE, Williams RJ. Learning representations by back-propagating errors. Nature. 1986;**323**(6088):533-536. DOI: 10.1038/323533a0

[38] Korotcov A, Tkachenko V, Russo DP, Ekins S. Comparison of deep learning with multiple machine learning methods and metrics using diverse drug discovery data sets. Molecular Pharmaceutics. 2017;**14**(12):4462-4475. DOI: 10.1021/acs.molpharmaceut.7b00578

[39] Whitehead TM, Irwin BWJ, Hunt P, Segall MD, Conduit GJ. Imputation of assay bioactivity data using deep learning. Journal of Chemical Information and Modeling. 2019;**59**(3):1197-1204. DOI: 10.1021/acs.jcim.8b00768

[40] Goodfellow I, Bengio Y, Courville A. Deep Learning. Cambridge, Massachusetts: The MIT Press; 2016. xxii. p. 775

[41] Torng W, Altman RB. 3D deep convolutional neural networks for amino acid environment similarity analysis. BMC Bioinformatics. 2017;**18**(1):302. DOI: 10.1186/s12859-017-1702-0

[42] Hinton GE, Salakhutdinov RR. Reducing the dimensionality of data with neural networks. Science. 2006;**313**(5786):504-507. DOI: 10.1126/science.1127647

[43] Xu Y, Lin K, Wang S, Wang L, Cai C, Song C, et al. Deep learning for molecular generation. Future Medicinal Chemistry. 2019;**11**(6):567-597. DOI: 10.4155/fmc-2018-0358

[44] Gomez-Bombarelli R, Wei JN, Duvenaud D, Hernandez-Lobato JM, Sanchez-Lengeling B, Sheberla D, et al. Automatic chemical design using a data-driven continuous representation of molecules. ACS Central Science. 2018;**4**(2):268-276. DOI: 10.1021/acscentsci.7b00572

[45] Jaeger S, Fulle S, Turk S. Mol2vec: Unsupervised machine learning approach with chemical intuition.

Journal of Chemical Information and Modeling. 2018;**58**(1):27-35. DOI: 10.1021/acs.jcim.7b00616

[46] Segler MHS, Kogej T, Tyrchan C, Waller MP. Generating focused molecule libraries for drug discovery with recurrent neural networks. ACS Central Science. 2018;**4**(1):120-131. DOI: 10.1021/acscentsci.7b00512

[47] Gagliardi F. Instance-based classifiers applied to medical databases: Diagnosis and knowledge extraction. Artificial Intelligence in Medicine. 2011;**52**(3):123-139. DOI: 10.1016/j.artmed.2011.04.002

[48] Asikainen AH, Ruuskanen J, Tuppurainen KA. Consensus kNN QSAR: A versatile method for predicting the estrogenic activity of organic compounds in silico. A comparative study with five estrogen receptors and a large, diverse set of ligands. Environmental Science & Technology. 2004;**38**(24):6724-6729. DOI: 10.1021/es049665h

[49] Bajorath J. Molecular similarity concepts for informatics applications. Methods in Molecular Biology. 2017;**1526**:231-245. DOI: 10.1007/978-1-4939-6613-4_13

[50] Lo YC, Senese S, Li CM, Hu Q, Huang Y, Damoiseaux R, et al. Large-scale chemical similarity networks for target profiling of compounds identified in cell-based chemical screens. PLoS Computational Biology. 2015;**11**(3):e1004153. DOI: 10.1371/journal.pcbi.1004153

[51] Kunkel C, Schober C, Oberhofer H, Reuter K. Knowledge discovery through chemical space networks: The case of organic electronics. Journal of Molecular Modeling. 2019;**25**(4):87. DOI: 10.1007/s00894-019-3950-6

[52] Keiser MJ, Roth BL, Armbruster BN, Ernsberger P, Irwin JJ, Shoichet BK. Relating protein pharmacology by ligand chemistry. Nature Biotechnology. 2007;**25**(2):197-206. DOI: 10.1038/nbt1284

[53] Louis B, Agrawal VK, Khadikar PV. Prediction of intrinsic solubility of generic drugs using MLR, ANN and SVM analyses. European Journal of Medicinal Chemistry. 2010;**45**(9):4018-4025. DOI: 10.1016/j.ejmech.2010.05.059

[54] Schneider G. Neural networks are useful tools for drug design. Neural Networks. 2000;**13**(1):15-16

[55] Asikainen A, Kolehmainen M, Ruuskanen J, Tuppurainen K. Structure-based classification of active and inactive estrogenic compounds by decision tree, LVQ and kNN methods. Chemosphere. 2006;**62**(4):658-673. DOI: 10.1016/j.chemosphere.2005.04.115

[56] Lagunin A, Zakharov A, Filimonov D, Poroikov V. QSAR Modelling of rat acute toxicity on the basis of PASS prediction. Molecular Informatics. 2011;**30**(2-3):241-250. DOI: 10.1002/minf.201000151

[57] Soufan O, Ba-Alawi W, Afeef M, Essack M, Kalnis P, Bajic VB. DRABAL: Novel method to mine large high-throughput screening assays using Bayesian active learning. Journal of Cheminformatics. 2016;**8**:64. DOI: 10.1186/s13321-016-0177-8

[58] Wilson DR, Martinez TR. Reduction techniques for instance-based learning algorithms. Machine Learning. 2000;**38**(3):257-286. DOI: 10.1023/A:1007626913721

[59] Svetnik V, Wang T, Tong C, Liaw A, Sheridan RP, Song Q. Boosting: An ensemble learning tool for compound classification and QSAR modeling. Journal of Chemical Information and Modeling. 2005;**45**(3):786-799. DOI: 10.1021/ci0500379

[60] Odziomek K, Rybinska A, Puzyn T. Unsupervised learning methods and similarity analysis in chemoinformatics. In: Leszczynski J, Kaczmarek-Kedziera A, Puzyn TG, Papadopoulos M, Reis HK, Shukla M, editors. Handbook of Computational Chemistry. Cham: Springer International Publishing; 2017. pp. 2095-2132

[61] Roy K, Pratim RP. Comparative chemometric modeling of cytochrome 3A4 inhibitory activity of structurally diverse compounds using stepwise MLR, FA-MLR, PLS, GFA, G/PLS and ANN techniques. European Journal of Medicinal Chemistry. 2009;**44**(7):2913-2922. DOI: 10.1016/j.ejmech.2008.12.004

[62] Bocker A, Derksen S, Schmidt E, Teckentrup A, Schneider G. A hierarchical clustering approach for large compound libraries. Journal of Chemical Information and Modeling. 2005;**45**(4):807-815. DOI: 10.1021/ci0500029

[63] Bocker A, Schneider G, Teckentrup A. NIPALSTREE: A new hierarchical clustering approach for large compound libraries and its application to virtual screening. Journal of Chemical Information and Modeling. 2006;**46**(6):2220-2229. DOI: 10.1021/ci050541d

[64] Zupan J, Gasteiger J, Zupan J. Neural Networks in Chemistry and Drug Design. 2nd ed. Weinheim; New York: Wiley-VCH; 1999. xxii. p. 380

[65] Schneider P, Tanrikulu Y, Schneider G. Self-organizing maps in drug discovery: Compound library design, scaffold-hopping, repurposing. Current Medicinal Chemistry. 2009;**16**(3):258-266

[66] Balasubramanian M, Schwartz EL. The isomap algorithm and topological stability. Science. 2002;**295**(5552):7. DOI: 10.1126/science.295.5552.7a

[67] L'Heureux PJ, Carreau J, Bengio Y, Delalleau O, Yue SY. Locally linear embedding for dimensionality reduction in QSAR. Journal of Computer-Aided Molecular Design. 2004;**18**(7-9):475-482

[68] Wallach I, Lilien R. The protein-small-molecule database, a non-redundant structural resource for the analysis of protein-ligand binding. Bioinformatics. 2009;**25**(5):615-620. DOI: 10.1093/bioinformatics/btp035

[69] Lo YC, Senese S, Damoiseaux R, Torres JZ. 3D chemical similarity networks for structure-based target prediction and scaffold hopping. ACS Chemical Biology. 2016;**11**(8):2244-2253. DOI: 10.1021/acschembio.6b00253

[70] Silver D, Huang A, Maddison CJ, Guez A, Sifre L, van den Driessche G, et al. Mastering the game of go with deep neural networks and tree search. Nature. 2016;**529**:484. DOI: 10.1038/nature16961. Available from: https://www.nature.com/articles/nature16961#supplementary-information

[71] Watkins CJCH, Dayan P. Q-learning. Machine Learning. 1992;**8**(3):279-292. DOI: 10.1007/BF00992698

[72] Mnih V, Kavukcuoglu K, Silver D, Rusu AA, Veness J, Bellemare MG, et al. Human-level control through deep reinforcement learning. Nature. 2015;**518**(7540):529-533. DOI: 10.1038/nature14236

[73] Popova M, Isayev O, Tropsha A. Deep reinforcement learning for de novo drug design. Science Advances. 2018;**4**(7):eaap7885. DOI: 10.1126/sciadv.aap7885

Chapter 11

Cell-Penetrating Peptides: A Challenge for Drug Delivery

Sonia Aroui and Abderraouf Kenani

Abstract

Cell-penetrating peptide (CPP) is a term that describes relatively short amphipathic and cationic peptides (7–30 amino acid residues) with rapid translocation across the cell membrane. They can be used to deliver molecular bioactive cargoes due to their efficacy in cellular internalization and also to their low cytotoxicity. In this review we provide an overview of the current approaches and describe the potential of CPP-based drug delivery systems and indicate their powerful promise for clinical efficacy.

Keywords: cell-penetrating peptides, drugs

1. Introduction

A novel approach to overcome cell membrane impermeability and to deliver a large variety of particles and macromolecules into cells has been recently emerged, which is called cell-penetrating peptides (CPPs), also known as protein transduction domains (PTDs) [1, 2]. CPPs are generally short (up to 30 amino acids in length) water-soluble, cationic, and/or amphipathic peptides which make them promising vectors for therapeutic delivery, leading to a considerable amount of research focused on the intracellular delivery of drugs [3–5]. There are two principal types of CPPs that have been utilized for this purpose: (i) cationic CPPs, composed of short sequence of amino acids (arginine, lysine, and histidine). The indicated amino acids give the cationic charge to the peptide and permit its interaction with anionic motifs on the plasma membrane by a receptor-independent mechanism. (ii) amphipathic peptides, which have lipophilic and hydrophilic tails that are responsible for a direct peptide translocation mechanism across the plasma membrane [6].

The most important characteristic of CPPs is that they are able to translocate the plasma membrane at low micromolar concentrations in vivo and in vitro without using any receptors and without causing any significant membrane damage [7, 8]. Other benefits of using CPPs for therapeutic delivery are the absence of toxicity as compared to other cytoplasmic delivery devices, such as liposomes, polymers, etc. [6]. The mechanism for the CPP-facilitated cellular uptake remains not clear and depends on cargo and cellular type [9]. Due to its high density of basic amino acid residues (Arg and Lys), the large charge at physiological pH excludes the passive diffusion of CPPs across the lipid bilayer. Furthermore, it seems that classical uptake mechanisms such as protein-based receptors and transporters are not involved. On the contrary, endocytosis was shown as a common uptake mechanism, but is controversial at the same time. For example, in a number of reports, CPP uptake was not inhibited at 4°C or in the presence of inhibitors of endocytosis; in contrast, a capture of CPPs in the endocytotic vesicles was observed when soluble heparin sulfate was added [9, 10]. Many other studies indicate that aggregation of the cell surface glycosaminoglycan heparan sulfate (HS) is an important element in the uptake mechanism [2]. The challenge of the strategy using CPPs should take into consideration the size, stability, nonspecific versus specific associations, and potency versus toxicity that all play an important role for the selection of delivery systems [5].

2. History and origin of CPPs

The CPPs are initially discovered in 1965 when it was observed that histones and cationic polyamines such as polylysine stimulate the uptake of albumin by tumor cells in culture. It was shown that the conjugation of polylysine to albumin and other proteins enhances their transport into cells. Moreover, a comparison study of different homopolymers of cationic amino acids demonstrates that medium-length polymers of arginine enter cells more effective than similar-length polymers composed of lysine, ornithine, or histidine [11]. In 1988, it was discovered that the human immunodeficiency virus type 1 (HIV-1) encoded trans-acting activator of transcription (Tat) peptide which also translocates cell membranes and gains intracellular mammalian cells [12, 13]. Covalently the conjugation of Tat peptide to proteins or fluorescent markers allowed these molecules to gain into the cell. A few years later, another discovery was followed when polycationic peptide of natural (VP22 and AntP) and synthetic origin (transportan) was used for the delivery of genes, proteins, small exogenous peptide, or even nanoparticles. Furthermore, it was demonstrated that small domains in these peptides are often responsible for cellular entry [14]. Thus, these translocation sequences could be shortened to a few amino acids in comparison with the first Tat peptide, without affecting cell penetration efficiency [13]. Since that time, the list of synthetic CPPs has increased sharply, and the number continues to rise (**Table 1**). In the last decade, another peptide was described named maurocalcine (MCa), a 33 amino acid residue peptide that has been isolated from the venom of the Tunisian chactid scorpion *Scorpio maurus palmatus.* It folds according to an "inhibitor cystine knot" (ICK) motif and contains three disulfide bridges connected by the following pattern: C1–C4, C2–C5, and C3– C6 [15]. MCa acts on ryanodine receptors resulting in pharmacological activation. These receptors are calcium channels located in the membrane of the endoplasmic reticulum. They control Ca^{2+} release from internal stores and therefore a large number of cell functions [16, 17].

This peptide possesses vector properties when coupled to fluorescent streptavidin. This complex was shown to enter various cell types within minutes and in all cell types tested, a common feature of CPPs. A variety of mutants of MCa were then designed in order to unravel the most active residues for its pharmacological and penetration activities (**Figure 1**) [18, 19].

3. Therapeutic applications of CPPs

3.1 CPP-cargo complex internalization mechanisms

Two distinct advances were shown to be used to bind CPPs to molecular cargoes. One process is non-covalently which connect CPP to its cargoes using electrostatic

Peptide	Sequence	Origin	Cargoes
Protein transduction domain			
Tat48-60	GRKKRRQRRRPPQ	VIH-1	ADN, peptide, PKC inhibitor
Pénétratin	RQIKIWFQNRRMKWKK	*Drosophila Antennapedia homeodomain*	HSP20 phosphopeptide
Chimeric peptides			
Transportan	GWTLNSAGYLLGKINLKALAALAKKIL	Galanin + Mastoparan	Protéine, PNA
Pep-1	KETWWETWWTEWSQPKKKRKV	Rich domain of tryptophan + *spacer* + domain derived from virus SV40-NLS sequence of T antigène	Enzyme
MPG	GALFLGFLGAAGSTMGAWSQPKKKRKV	Hydrophobic motif derived from HIV-1 gp41 + *linker* + *domain derived* from virus SV40-NLS sequence of T antigène	siARN, oligo-nucléotides
CADY	GLWRALWRLLRSLWRLLWRA	Dérived from PPTG11, variant of JTS1 fusion protéin	siARN
Peptide models			
(Arg)x	(RRRRR)X	Synthetic peptide	siARN, Cyclosporine A
MAP	KLALKLALKALKAALKA	Synthetic peptide	Natural CPPs
Natural CPP			
Maurocalcine	GDCLPHLKLCKENKDCCSKKCKRRGTNIEKRCR	*Scorpio maurus palmatus*	Doxorubicin

Table 1.
Examples of four classes of CPPs and delivered cargoes. The list of cargo is not exhaustive and given for illustration. X = 7, 8, or 9 arginine residues.

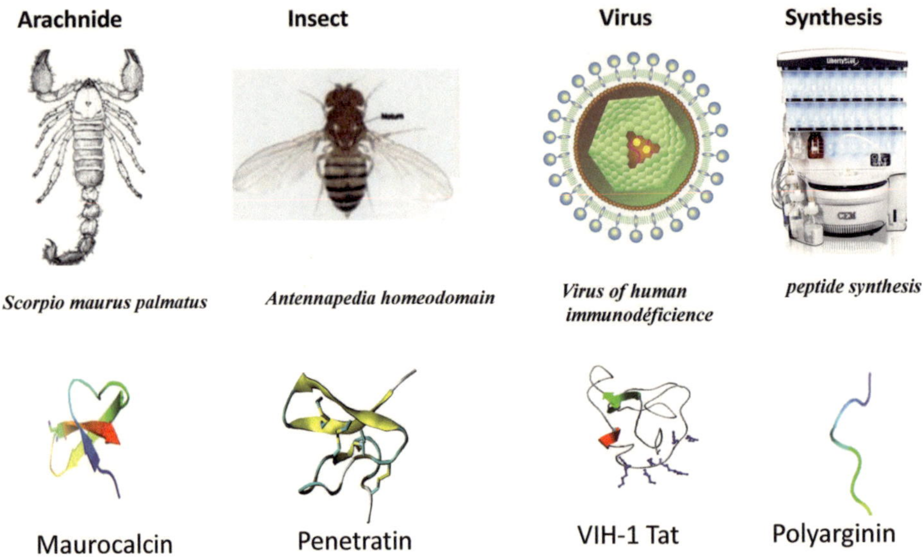

Figure 1.
Example of origin of four CPPs: Maurocalcine, penetratin, tat, and polyarginine. Maurocalcine, penetratin, and tat are derived from natural sequences, but polyarginine was produced by de novo conception in order to obtain a good cellular penetration.

interactions, such as MPG and Pep-1, amphipathic peptides carriers, which link to cargoes beyond any cross linking or chemical changes [20]. The second approach is more frequent and uses a covalent relation between the two compounds. This means has been widely used by different teams and has demonstrated positive advances, especially with TAT, penetratin, or polyarginines [21].

Various mechanisms for CPP internalization have been suggested, but the exact one is still not well known. Yet, many data approve that the energy-dependent tool (endocytosis) and the energy-independent mechanism (direct translocation) or both are involved in the cellular uptake progress [22].

For direct penetration, various mechanisms have been described: the carpet-like model (membrane destabilization) [23] and the pore formation model (barrel-stave) [24]. Positively charged CPPs interact with negatively charged membrane components like phospholipid bilayer or heparan sulfate. Such interaction is dwelling on the first stage of all of these mechanisms, followed by destabilization of the membrane and finished by crossing of the CPP on the lipid membrane.

For endocytosis transduction or cellular digestion, pinocytosis, phagocytosis, and receptor-mediated endocytosis have been reported [25, 26]. A sum-up of CPP transduction systems is shown in **Figure 2**. In pinocytosis, the plasma membrane absorbs solutes, while in phagocytosis it takes great particles. In clathrin-mediated endocytosis, clathrin and also caveolin, which are receptor-mediated endocytosis and cover the intracellular part of the biomembranes, possess a key role in the uptake mechanism. These protein structures are pivotal for the membrane invagination and for the construction of the vesicles after bounding the extracellular molecule to the membrane receptor. Clathrin has a great diameter in comparison with caveolin-coated vesicles and was also considered as a selective route for the translocation of compounds into cells through specific receptors on the surface of the cell [27].

Many determinants influence the internalization process, such as the nature of CPP or the cell type, the cargo, and the experimental conditions (temperature and pH) [22].

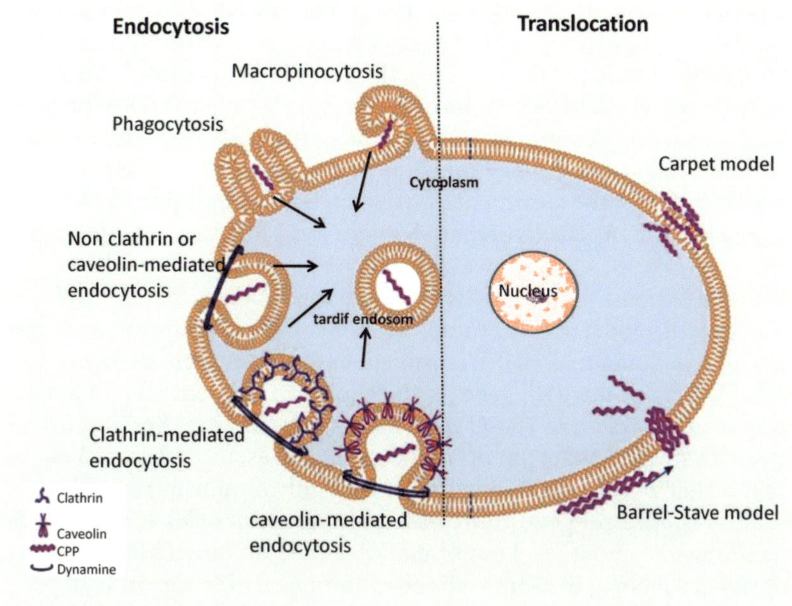

Figure 2.
CPP translocation mechanisms.

3.2 Delivery of chemotherapeutic agents

Chemotherapy used for treatment of cancer has a lot of defects because of the toxicity of the drugs to normal healthy cells and also to resistance developed by tumor cells to the anticancer drug [28]. The major inconveniences with used cancer chemotherapy are the absence of specificity target to tumor cells and thus poor antitumor effect. The challenge in cancer therapy is to know how to deliver a drug intact to the cytosol of every cancer cell, sparing healthy cells.

It was shown that polyarginines carry cargoes that exceed 500 Da by molecular electroporation across the cell membrane which may solve part of the drug delivery problem [29]. However, the use of well-chosen linkers and anions can help target cancer cells and contribute to successful conjugation process. For example, the CXC chemokine receptor 4 (CXCR4) is overexpressed in different types of cancer, including prostate, breast, colon, and small-cell lung cancer. Snyder et al. linked the CXCR4 receptor ligand, DV3, to two transducible anticancer peptides: a p53-activating peptide (DV3-TATp53C′) and a cyclin-dependent kinase 2 antagonist peptide (DV3-TAT-RxL). Treatment of tumor cells expressing the CXCR4 receptor with either the DV3-TATp53C′ or DV3-TAT-RxL targeted peptides resulted in an enhancement of tumor cell killing compared with treatment with nontargeted parental peptides [30]. Furthermore, hypoxia-inducible factor-1 (HIF), the transcription factor central to oxygen homeostasis, is regulated via the oxygen-dependent degradation domains (ODD) of its α isoforms (HIFα). The amino- and carboxyl-terminal sequences of ODD (NODD and CODD) were fused to TAT and injected into sponges implanted subcutaneously (s.c.) in mice by William et al. They demonstrated that this injection causes a markedly accelerated local angiogenic response and induction of glucose transporter-1 gene expression, thus opening additional therapeutic avenues for ischemic diseases [31].

In some cancer cells, such as melanoma (common eye cancers in adults), p53 seems to be inhibited by overexpression of HDM2. A transducible peptide that inhibits HDM2 and Bcl-2 for their ability to induce tumor-specific apoptosis in these cells was tested [30]. In this study, it was demonstrated that the anti-Bcl-2 peptide induced apoptosis in tumor cells but also caused variable levels of toxicity in normal cells and tissues. On the contrary, the anti-HDM2 peptide induced apoptosis in tumor cells, with little effect on normal cells in a therapeutic dose range. This peptide also caused regression of retinoblastoma in rabbit eyes, with minimal damage to normal ocular tissues. They conclude that the inhibition of HDM2 may be a promising strategy for the treatment of uveal melanoma and retinoblastoma, and that strategy may be an effective technology for local delivery of anticancer therapy to the eye.

Most of the patients with sporadic renal cell carcinomas (RCCs) exhibit mutation of the Hippel-Lindau (VHL) tumor suppressor gene. Conjugation of the protein transduction domain of HIV-TAT protein to the amino acid sequence (104–123) in the beta-domain of the VHL gene product (pVHL) arrested and then reduced proliferation and invasion of 786-O renal cancer cells in vitro. Besides, daily i.p. injections with the conjugate put off and, in some cases, caused partial regression of renal tumors that were implanted in the dorsal flank of nude mice [32].

The tumor suppressor gene *p16INK4A*, an inhibitor of cdk3 4, is often inactivated via intragenic mutation, homozygous deletion, and methylation-associated transcriptional silencing in a large number of human cancers, mainly in pancreatic cancer. Treated animals with the p16-derived synthetic peptide coupled with the Antennapedia carrier sequence, in which we designated as Trojan p16 peptide, showed reduced AsPC-1 and BxPC-3 s.c. tumors, respectively. Thus, we conclude that Trojan p16 peptide system, a gene-oriented peptide coupled with a peptide vector, functions for experimental pancreatic cancer therapy [33].

Recently, it was shown by Sonia et al. that coupling doxorubicin (Dox) to three cell-penetrating peptides Tat, penetratin, and maurocalcine (Dox-CPPs) is a good strategy to overcome Dox resistance in MDA-MB 231 breast cancer cells and CHO cells (**Figure 3**) [3, 34]. We also reported that all conjugates are able to promote cell apoptosis in the breast cancer-resistant cells MDA-MB 231 at lesser concentration needed for Dox alone. Indeed, apoptosis death was shown to be correlated with ladder-internucleosomal degradation, chromatin contraction, caspase activation, Bad and Bax activation by oligomerization on the mitochondrial membrane, and liberation of cytochrome c. Despite the effective Bcl-2 overexpression in apoptosis induced by the Dox alone, such potency was shown to be insufficient in case of Dox-CPP-triggered cell apoptotic death. Otherwise, these results suggest that there are other apoptotic signaling pathways, independent of mitochondrial one, which are implicated in Dox-CPP apoptosis. Moreover, greater effectiveness of Dox when coupled to CPPs is not due only to its higher accumulation on the cells but also to the incitement of other signaling pathways. These pathways include death receptors and activation of the JNK pathway [4, 35].

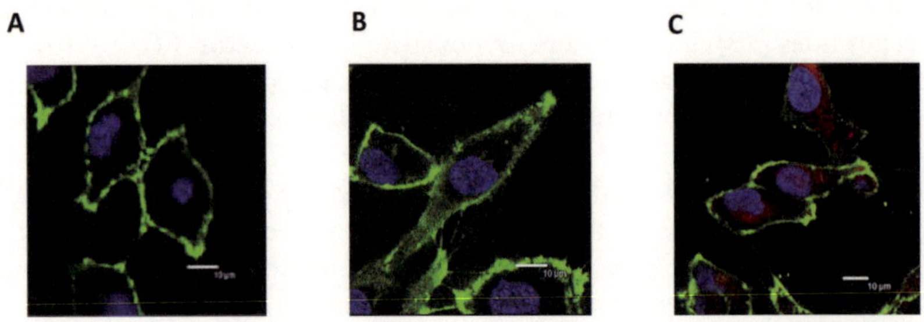

Figure 3.
Cellular internalization of Dox by MCa. MDA-MB231 cells treated with (a) RPMI, (B) Dox alone (red), and (C) Dox coupled to Dox at the same concentration (red).

Another study led by Leslie Walker et al. showed that conjugated Dox to both ELP and SynB1 prevents tumor development in mice. In fact, conjugation of Dox to SynB1-ELP was more efficient in tumor inhibition under hyperthermic condition than Dox alone, which was twofold higher. Such conception was considered hopeful peptide candidates for drug delivery [36]. The anticancer activity of Dox was also enhanced when constructed a drug delivery system by developing 25 nm gold nanospheres (GNSs) conjugated to four α-helical CPPs [37].

A thermally sensitive quantum dot that exhibits an "on-demand" cellular uptake behavior via temperature-induced "shielding/deshielding" of CPP on the surface was synthesized. Poly(N-isopropylacrylamide) (PNIPAAm) and CPP were biotinylated at their terminal ends and co-immobilized onto the surface of streptavidin-coated quantum dots (QDs-Strep) through biotin-streptavidin interaction. Namely, under a lower critical solution temperature (LCST), the hydrated PNIPAAm chains blocked CPP cellular uptake. This effect was broken down when the LCST was raised to allow CPP moieties to be exposed on the cell surface, leading to QD cellular uptake.

Additionally, the "shielding/deshielding" temperature of CPP was also used for siRNA delivery system. Biotinylated siRNA was coupled to the surface of TSQDs. Indeed, the amount of corresponding gene silencing was increased due to the surface exposure of CPP within a rising temperature above the LCST [38].

4. Optimization methods for CPP-mediated cancer therapy and diagnosis

Over the last decade, a great attention has been assigned to the importance of CPP on drug transportation of bioactive molecules in various preclinical studies. In fact, novel computational basics have been made in order to develop knowledge on CPPs [39].

Previously, different researchers have developed a few in silico algorithm approaches for CPP prediction (CPPpred) and screening to facilitate throughput CPP-based research. The in silico screening/prediction methods aimed on the use of scales of chemical characteristic, such as z-descriptors [40, 41]. It is generally followed by experimental validation to make it reliable with less cost and time-consuming approach. Later on, other CPP prediction applied neural network (NN) strategies were developed and consist on introducing an N-to-1 NN. The network proceeds by a sequence of 5 to 30 amino acids in length, as input, and gives a prediction of how probably each peptide is to be cell penetrating [42]. This CPPpred offers an advantage since it was developed with repetition-reduced training and test sets.

Over the years, the commitment therapeutic importance of CPPs motivated other teams to develop the first version of CPP database, i.e., CPPsite which supports broad information on the promising use of CPPs [43]. The CPPsite manually created database of 843 experimentally described CPPs. Each consulting gives us data of the peptide involving peptide sequence, peptide name, nature of peptide, origin, chirality, uptake efficiency, subcellular localization, etc. A deep area of user-friendly tools has been integrated in this database like analyzing and browsing tools. Moreover, they have introduced other informations concerning peptide sequences such as secondary/tertiary structure and physicochemical properties of peptides.

This database version was then developed and updated as a CPPsite 2.0 and holds 1855 entries, including 1012 recent new entries [44]. The renovated version contains further data concerning chemically modified CPPs used on the in vivo model. In addition to other informations on delivered cargoes by CPPs (proteins, molecules, nanoparticles, DNA, RNA, etc.), secondary and tertiary structures of natural and chemical CPPs (including CPP with D-amino acids) were also predicted in view of their important role in the functionality of CPPs and stored in the database. Numerous tools for information browse and analysis are combined in this database and considered as a useful resource since it is compatible for all users, including smartphone and tablet.

CPP prediction sites are a promising assist to the researchers to design cell penetrating peptide, as well as making different modification and to investigate their effect on cell penetration potency [45].

5. Conclusion

The progressive and continuous application of CPPs shows that they are efficient delivery vectors. Because of the need to ameliorate the drug delivery, a great number of CPP-based applications are still drawing the attention of researchers.

In this review, the current tendency in drug delivery by CPPs is summed up. Conjugation with CPP increases cell-surface affinity and eventual cellular uptake of bioactive molecules.

Author details

Sonia Aroui* and Abderraouf Kenani
Unité de Recherche UR 12ES08 "Signalisation Cellulaire et Pathologies", Faculté de Médecine de Monastir, Monastir, Tunisie

*Address all correspondence to: sonia_aroui2002@yahoo.fr

References

[1] El-Sayed A, Futaki S, Harashima H. The AAPS Journal. 2009;**11**:13-22

[2] Moon JI, Han MJ, Yu SH, Lee EH, Kim SM, Han K, et al. Enhanced delivery of protein fused to cell penetrating peptides to mammalian cells. BMB Reports. May 2019;**52**(5):324-329

[3] Aroui S, Ram N, Appaix F, Ronjat M, Kenani A, Pirollet F, et al. Maurocalcine as a non toxic drug carrier overcomes doxorubicin resistance in the cancer cell line MDA-MB 231. Pharmaceutical Research. 2008;**10**:9782-9801

[4] Aroui S, Brahim S, De Waard M, Bréard J, Kenani A. Efficient induction of apoptosis by doxorubicin coupled to cell-penetrating peptides compared to unconjugated doxorubicin in the human breast cancer cell line MDA-MB 231. Cancer Letters. 2009;**285**:28-38

[5] Lönn P, Dowdy SF. Cationic PTD/CPP-mediated macromolecular delivery: Charging into the cell. Expert Opinion on Drug Delivery. 2015;**12**(10):1627-1636

[6] Schroeder JA, Bitler BG. Anti-cancer therapies that utilize cell penetrating peptides. Recent Patents on Anti-Cancer Drug Discovery. 2010;**5**:1-10

[7] Jarver P, Langel U. Cell-penetrating peptides-a brief introduction. Biochimica et Biophysica Acta. 2006;**1758**:260-263

[8] Lehto T, Kurrikoff K, Langel U. Cell-penetrating peptides for the delivery of nucleic acids. Expert Opinion on Drug Delivery. 2012;**9**:823-836

[9] Ram N, Aroui S, Jaumain E, Bichraoui H, Mabrouk K, Ronjat M, et al. Direct peptide interaction with surface glycosaminoglycans contributes to the cell penetration of maurocalcine. The Journal of Biological Chemistry. 2008;**29**:24274-24284

[10] Wu X, Gehring W. Cellular uptake of the *Antennapedia homeodomain* polypeptide by macropinocytosis. Biochemical and Biophysical Research Communications. 2013;**13**:1-1

[11] Ziegler A, Nervi P, Durrenberger M, Seelig J. The cationic cell-penetrating peptide CPP TAT derived from the HIV-1 protein tat is rapidly transported into living fibroblasts: Optical, biophysical and metabolic evidence. Biochemistry. 2005;**44**:138-148

[12] Green M, Loewenstein PM. Autonomous functional domains of chemically synthesized human immunodeficiency virus tat trans-activator protein. Cell. 1988;**55**:1179-1188

[13] Vives E, Brodin P, Lebleu B. A truncated HIV-1 tat protein basic domain rapidly translocates through the plasma membrane and accumulates in the cell nucleus. The Journal of Biological Chemistry. 1997;**272**:16010-16017

[14] Joliot A, Prochiantz A. Transduction peptides: From technology to physiology. Nature Cell Biology. 2004;**6**:189-196

[15] Fajloun Z, Kharrat R, Chen L, Lecomte C, Di Luccio E, Bichet D, et al. Chemical synthesis and characterization of maurocalcine, a scorpion toxin that activates Ca(2+) release channel/ryanodine receptors. FEBS Letters. 2000;**469**:179-185

[16] Esteve E, Smida-Rezgui S, Sarkozi S, Szegedi C, Regaya I, Chen L, et al. Critical amino acid residues determine the binding affinity and the Ca2+ release efficacy of maurocalcine in skeletal muscle cells. The Journal of Biological Chemistry. 2003;**278**:37822-37831

[17] Esteve E, Mabrouk K, Dupuis A, Smida-Rezgui S, Altafaj X, Grunwald D, et al. Transduction of the scorpion toxin maurocalcine into cells. Evidence that the toxin crosses the plasma membrane. The Journal of Biological Chemistry. 2005;**280**:12833-12839

[18] Mabrouk K, Ram N, Boisseau S, Strappazzon F, Rehaim A, Sadoul R, et al. Critical amino acid residues of maurocalcine involved in pharmacology, lipid interaction and cell penetration. Biochimica et Biophysica Acta. 2007;**1768**:2528-2540

[19] Tisseyre C, Bahembera E, Dardevet L, Sabatier JM, Ronjat M, De Waard M. Cell penetration properties of a highly efficient mini maurocalcine peptide. Pharmaceuticals. 2013;**18**:320-339

[20] Feni L, Neundorf I. The current role of cell-penetrating peptides in cancer therapy. Advances in Experimental Medicine and Biology. 2017;**1030**:279-295

[21] Nischan N, Herce HD, Natale F, Bohlke N, Budisa N, Cardoso MC, et al. Covalent attachment of cyclic TAT peptides to GFP results in protein delivery into live cells with immediate bioavailability. Angewandte Chemie (International Ed. in English). 2015;**54**:1950-1953

[22] Madani F, Lindberg S, Langel Ü, Futaki S, Gräslund A. Mechanisms of cellular uptake of cell-penetrating peptides. Biophysical Journal. 2011;**2011**:414729

[23] Matsuzaki K, Sugishita K, Miyajima K. Interactions of an antimicrobial peptide, magainin 2, with lipopolysaccharide-containing liposomes as a model for outer membranes of gram-negative bacteria. FEBS Letters. 1999;**449**:221-224

[24] Kawamoto S, Takasu M, Miyakawa T, Morikawa R, Oda T, Futaki S, et al. Inverted micelle formation of cell-penetrating peptide studied by coarse-grained simulation: Importance of attractive force between cell-penetrating peptides and lipid head group. The Journal of Chemical Physics. 2011;**134**:095103

[25] Mayor S, Pagano RE. Pathways of clathrin-independent endocytosis. Nature Reviews. Molecular Cell Biology. 2007;**8**:603-612

[26] Ferrari A, Pellegrini V, Arcangeli C, Fittipaldi A, Giacca M, Beltram F.Caveolae-mediated internalization of extracellular HIV-1 tat fusion proteins visualized in real time. Molecular Therapy. 2003;**8**:284-294

[27] Habault J, Poyet JL. Recent advances in cell penetrating peptide-based anticancer therapies. Molecules. 2019;**24**:927

[28] Mae M, Myrberg H, EI-Andaloussi S, Langel U. Design of a tumor homing cell-penetrating peptide for drug delivery. International Journal of Peptide Research and Therapeutics. 2009;**15**:11-15

[29] Cahill K. Cell-penetrating peptides, electroporation and drug delivery. IET Systems Biology. 2010;**4**:367-378

[30] Harbour JW, Worley L, Ma D, Cohen M. Transducible peptide therapy for uveal melanoma and retinoblastoma. Archives of Ophthalmology. 2002;**120**:1341-1346

[31] Snyder E, Saenz C, Denicourt C, Meade BR, Cui XS, Kaplan IM, et al. Enhanced targeting and killing of tumor cells expressing the CXC chemokine receptor 4 by transducible anti-cancer peptides. Cancer Research. 2005;**65**:10646-10650

[32] Datta K, Sundberg C, Karumanchi SA, Mukhopadhyay D. The 104-123 amino acid sequence of the β-domain of von Hippel-Lindau gene product is sufficient to inhibit renal tumor growth and invasion. Cancer Research. 2001;**61**:1768-1775

[33] Hosotani R, Miyamoto Y, Fujimoto K, Doi R, Otaka A, Fujii N, et al. Trojan p16 peptide suppresses pancreatic cancer growth and prolongs survival in mice. Clinical Cancer Research. 2002;**8**:1271-1276

[34] Aroui S, Mili D, Brahim S, De Waard M, Kenani A. Doxorubicin coupled to penetratin promotes apoptosis in CHO cells by a mechanism involving c-Jun NH2-terminal kinase. Biochemical and Biophysical Research Communications. 2010;**396**:908-914

[35] Aroui S, Brahim S, Hamelin J, De Waard M, Bréard J, Kenani A. Conjugation of doxorubicin to cell penetrating peptides sensitizes human breast MDA-MB 231 cancer cells to endogenous TRAIL-induced apoptosis. Apoptosis. 2009;**11**:1352-1365

[36] Walker L, Perkins E, Kratz F, Raucher D. Cell penetrating peptides fused to a thermally targeted biopolymer drug carrier improve the delivery and antitumor efficacy of an acid-sensitive doxorubicin derivative. International Journal of Pharmaceutics. 2012;**436**:825-832

[37] Park H, Tsutsumi H, Mihara H. Cell-selective intracellular drug delivery using doxorubicin and α-helical peptides conjugated to gold nanoparticles. Biomaterials. 2014;**10**:3480-3487

[38] Kim C, Lee Y, Kim JS, Jeong JH, Park TG. Thermally triggered cellular uptake of quantum dots immobilized with poly(N-isopropylacrylamide) and cell penetrating peptide. Langmuir. 2010;**26**:14965-14969

[39] Damiati SA et al. Novel machine learning application for prediction of membrane insertion potential of cell-penetrating peptides. International Journal of Pharmaceutics. 2019;**567**(15):118453

[40] Sandberg M et al. New chemical descriptors relevant for the design of biologically active peptides. A multivariate characterization of 87 amino acids. Journal of Medicinal Chemistry. 1998;**41**:2481-2491

[41] Ha¨llbrink M et al. Prediction of cell-penetrating peptides. International Journal of Peptide Research and Therapeutics. 2005;**11**:249-259

[42] Holton TA, Pollastri G, Shields DC, Mooney C. CPPpred: Prediction of cell penetrating peptides. Bioinformatics. 2013;**29**(23):3094-3096

[43] Gautam A, Singh H, Tyagi A, Chaudhary K, Kumar R, Kapoor P, et al. CPPsite: A curated database of cell penetrating peptides. Database: The Journal of Biological Databases and Curation. 2012;**2012**:bas015

[44] Agrawal P, Bhalla S, Usmani SS, Singh S, Chaudhary K, Raghava GP, et al. CPPsite 2.0: A repository of experimentally validated cell penetrating peptides. Nucleic Acids Research. 2016;**44**(D1):D1098-D1103

[45] Kang Z et al. The rational design of cell-penetrating peptides for application in delivery systems. Peptides. 2019;**121**:170149